Agricultural Equipment Operations

Richard Skiba

AFTER MIDNIGHT
PUBLISHING

Skiba, Richard (author)

Agricultural Equipment Operations

ISBN 978-1-7635254-1-2 (paperback) 978-1-7635254-2-9 (eBook) 978-1-7635254-3-6 (Hardcover)

Non-fiction

Contents

Preface

This book covers the operation of a selective range of agricultural equipment, specifically:

- Tractors
- Cultivators
- Windrowers
- Disc Harrows
- Seeders/ Seed drills/Planters
- Sprayers (including pesticides and fertilizers)
- Combine harvesters
- Grain carts
- Balers (for hay and straw) and Bale Wrappers
- Mowers and Slashers
- Spreaders – Fertilizer and Manure
- Quad Bikes

For each of these, the uses, key components, operating principles, preparation for operations, operational practices, safe operation and finalising operations is covered.

The agricultural equipment information provided within this book is intended to be general in nature and may not encompass all aspects of its operation. It is important to note that each item of plant or equipment has its own specific characteristics and operational requirements that may vary. Agricultural equipment operators are strongly advised to consult the manufacturer's guides and manuals prior to the operation of any equipment to ensure compliance with safety standards and operational procedures.

Furthermore, it is crucial to acknowledge that operations and terminology can differ across jurisdictions. Agricultural equipment operators should be aware that regulations and guidelines pertaining to equipment usage may vary depending on the location. Therefore, it is essential for equipment operators to familiarize themselves with the applicable laws, regulations, and standards in their respective jurisdictions.

Additionally, agricultural equipment operators are urged to review workplace policies and procedures before operating any equipment. Workplace-specific protocols may exist to address unique hazards and safety considerations, which must be adhered to for safe operations.

Moreover, it is important to recognize that in many jurisdictions, operational licensing requirements apply. Agricultural equipment operators are responsible for ensuring that they meet all jurisdictional legislative requirements relevant to their sites of practice. This may include obtaining appropriate licenses, certifications, or permits to operate equipment legally and safely within their jurisdiction.

Sample load charts, specifications, interpretations and calculations are used throughout this book for demonstration purposes only and should not be taken to be used in any other manner. Every equipment model is accompanied by its own distinct operational charts and characteristics, which may vary depending on the equipment's configurations and rated capacity and is supplied by the equipment's manufacturer. They are not portable from one model to another, and operators must always ensure they are referring to documentation relevant to the plant they are operating.

While efforts have been made to provide accurate and informative equipment operation information, users are reminded of the need for due diligence and compliance with applicable regulations, manufacturer guidelines, workplace policies, and licensing requirements to ensure safe and lawful equipment operations.

Chapter One

Introduction

Agricultural equipment refers to a wide range of machinery, tools, and devices specifically designed and utilized for various tasks in agricultural production and farming operations. These tools are essential for cultivating, planting, harvesting, and processing crops, as well as managing livestock and maintaining agricultural infrastructure. Agricultural equipment encompasses both powered and non-powered machinery, including tractors, plows, planters, harvesters, irrigation systems, spraying equipment, livestock handling facilities, and storage structures. The use of agricultural equipment has significantly improved efficiency, productivity, and output in modern agriculture, enabling farmers to manage larger areas of land and meet the demands of global food production.

Agricultural tools and equipment play vital roles in modern farming operations, each serving specific purposes to enhance efficiency and productivity on farms. Tractors, for instance, are versatile vehicles indispensable for a wide range of tasks including plowing, tilling, planting, harvesting, and hauling. They form the backbone of agricultural operations, providing power and mobility to accomplish various farm tasks.

Figure 1: A New Holland T 4.85 +2.1p tractor pulling a cultipacker over a dry field in Gåseberg. W.carter, CC0, via Wikimedia Commons.

Cultivators, on the other hand, are specialized implements used to break up soil and remove weeds between rows of crops. By promoting soil aeration and water infiltration, cultivators contribute to optimal growing conditions for crops. Seeders and planters are essential for accurate and efficient planting of seeds, ensuring optimal seed placement at specific depths and spacing, thereby maximizing crop yields.

Figure 2: Fendt 412 Vario with Horsch Terrano 3 FX cultivator on a field. Reinhold Möller, CC BY-SA 4.0 , via Wikimedia Commons.

Sprayers, including those used for pesticides and fertilizers, play a crucial role in crop protection and nutrient supplementation. They are utilized to apply chemicals and nutrients to crops, ensuring effective pest control and soil fertility management. Combine harvesters are large machines designed for harvesting grain crops such as wheat, barley, and corn, performing multiple tasks including cutting, threshing, and cleaning in a single pass.

Balers are employed to compress cut hay or straw into dense bales for storage, transport, and feeding livestock, while mowers are utilized for cutting grass, hay, and other crops for silage, haymaking, or pasture management. Tillers, grain carts, and grain augers aid in soil preparation, harvesting, and grain transportation, respectively, contributing to overall harvesting efficiency.

Fertilizer spreaders and manure spreaders are essential for evenly distributing fertilizers and animal manure onto fields, replenishing soil nutrients and improving crop yields. Seed drills ensure precise and uniform seed placement into the soil, optimizing seedling emergence and growth. Windrowers arrange crops like hay or small grains into windrows for drying or harvesting.

Bale wrappers preserve hay or silage bales by wrapping them in plastic film, maintaining moisture and quality during storage. Disc harrows break up and level soil, preparing seedbeds for planting by incorporating crop residues. Row crop cultivators remove weeds and aerate soil between rows of crops, promoting crop health and growth.

Figure 3: Farmer on quad bike. Peter Barr, CC BY-SA 2.0 , via Wikimedia Commons.

Lastly, quad bikes, also known as all-terrain vehicles (ATVs), serve multiple purposes on farms including transportation, herding livestock, and accessing remote areas of the property. Each of these tools and equipment is integral to modern agricultural practices, contributing to efficient and productive farming operations.

Operating agricultural equipment comes with a range of hazards and risks that must be carefully managed to prevent injuries, accidents, and damage. Each type of equipment used on farms presents its own set of hazards and associated risks:

Tractors, for instance, pose hazards such as rollovers, entanglement in moving parts, crush injuries, and falls from heights. Risks include navigating uneven terrain, improper use of attachments, and inadequate operator training.

Cultivators present hazards like entanglement in rotating parts, struck-by injuries, and slips and falls. Risks include contact with underground utilities, navigating uneven terrain, and improper maintenance.

Seeders and planters are associated with hazards such as entanglement in moving parts and crush injuries during loading/unloading. Risks include improper calibration, seed spillage, and exposure to dust and chemicals.

Sprayers, including those used for pesticides and fertilizers, pose hazards such as chemical exposure, inhalation of fumes, and skin and eye injuries. Risks include improper handling and mixing, leaks or spills, and inadequate personal protective equipment (PPE).

Combine harvesters present hazards like entrapment in machinery, falls from elevated platforms, and struck-by injuries. Risks include malfunctioning equipment, inadequate visibility, and navigating uneven terrain.

Balers for hay and straw are associated with hazards such as entanglement in moving parts and crush injuries during the baling process. Risks include blockages in machinery, improper maintenance, and inadequate guarding.

Figure 4: Small Squre Baler with Norden Mfg Bale Accumulator. Glendon Kuhns, CC0, via Wikimedia Commons.

Mowers used for cutting grass and crops pose hazards like contact with blades, thrown objects, and noise exposure. Risks include slips and falls, collision with obstacles, and inadequate safety guards.

Tillers present hazards such as entanglement in rotating parts, struck-by injuries, and noise exposure. Risks include improper operation, inadequate maintenance, and contact with buried objects.

Grain carts are associated with hazards like falls from elevated platforms, struck-by injuries, and equipment rollovers. Risks include navigating uneven terrain, overloading, and improper hitching to tractors.

Each equipment type carries specific hazards and risks that must be managed through proper safety protocols, adequate operator training, regular maintenance, and the use of appropriate personal protective equipment (PPE) when necessary. These measures help ensure the safe and efficient operation of agricultural equipment on farms.

Understanding the operational principles of agricultural equipment is crucial for operators due to several key reasons. Firstly, safety is paramount in preventing accidents and injuries. Operators must grasp how to utilize safety features effectively, identify potential hazards, and react appropriately to unforeseen circumstances to ensure a safe working environment.

Secondly, operational efficiency is essential for maximizing productivity and minimizing downtime caused by malfunctions or errors. With a comprehensive understanding of equipment operation, operators can optimize performance and ensure tasks are completed efficiently.

Moreover, knowledge of equipment operation facilitates proactive maintenance practices. Operators can identify signs of wear and address maintenance needs promptly, thereby reducing the risk of breakdowns and extending the lifespan of the equipment.

Furthermore, familiarity with operational principles enables operators to perform tasks accurately and effectively, resulting in higher-quality work. Whether planting seeds, applying pesticides, or baling hay, precise execution enhances overall productivity and yield.

In addition to improving efficiency and quality, effective equipment operation leads to cost savings. Operators who understand how to use equipment efficiently can minimize fuel consumption and other operating expenses, thereby maximizing profitability.

Moreover, proper equipment operation can have positive environmental implications. By minimizing chemical runoff, soil compaction, and energy consumption, operators can help reduce environmental impacts and promote sustainability in agricultural practices.

Lastly, adherence to regulatory requirements is crucial for legal compliance. Many agricultural operations are subject to regulations governing equipment use and environmental practices. Understanding operational principles helps operators comply with these regulations, avoiding fines and penalties.

Operators with a thorough understanding of the operational principles of agricultural equipment are better equipped to work safely, efficiently, and responsibly. By prioritizing safety, optimizing efficiency, and adhering to regulatory standards, operators contribute to the success and sustainability of farming operations.

Chapter Two

Tractors

A tractor is a powerful motorized vehicle designed primarily for pulling or pushing agricultural machinery or trailers. Tractors are widely used in farming and construction industries for various tasks such as plowing, tilling, planting, and harvesting crops, as well as for hauling materials and equipment.

Tractors typically have large rear wheels with deep treads for traction in various terrains and may also have smaller front wheels for steering. They are equipped with a powerful engine, usually diesel-powered, to provide the necessary torque and horsepower to operate attached implements. Tractors often have a sturdy frame and chassis to withstand the stresses of heavy-duty work.

Tractors come in different sizes and configurations depending on the specific tasks they are intended for and the scale of operation, ranging from compact utility tractors used in small farms and landscaping to large, high-horsepower models used in large-scale agricultural operations. Additionally, modern tractors may be equipped with advanced technology such as GPS guidance systems and automated controls to improve efficiency and precision in farming operations.

Figure 5: A modern 4-wheel drive farm tractor. Solitude, CC BY-SA 2.0, via Wikimedia Commons.

The operational principles of a tractor encompass its functioning and capabilities in various agricultural and industrial tasks. Here's an overview:

1. Engine Power: Tractors are powered by internal combustion engines, typically fuelled by diesel or gasoline. The engine provides the necessary power to operate the tractor and drive its various components.

2. Transmission: Tractors utilize a transmission system to transfer power from the engine to the wheels. This system allows the operator to control the speed and direction of the tractor, typically through gears or hydrostatic transmissions.

3. Steering: Tractors employ a steering mechanism to navigate through fields and other terrain. Steering can be accomplished through various methods, including manual steering, power steering, or even advanced systems like GPS-guided steering.

4. Hydraulics: Hydraulic systems are integral to the operation of tractors, providing power for implements such as loaders, plows, and mowers. Hydraulics allow for precise control over the movement and operation of these implements, enhancing efficiency and productivity.

5. Power Take-Off (PTO): Tractors often feature a power take-off shaft that transfers power from the engine to external implements. This enables the tractor to drive machinery such as pumps, generators, and hay balers, expanding its versatility in agricultural and industrial applications.

6. Three-Point Hitch: Many tractors are equipped with a three-point hitch system, which allows implements to be easily attached and adjusted. This system consists of two lower lift arms and an upper link, providing stability and control over attached implements during operation.

7. Traction and Stability: Tractors are designed with features to ensure traction and stability, crucial for navigating varied terrain and carrying out tasks safely. This includes features like differential locks, tyre options for different conditions, and weight distribution systems.

8. Operator Controls: Tractors are equipped with control panels or consoles that allow operators to manage various functions such as engine speed, implement operation, and hydraulic controls. Modern tractors may also incorporate advanced technology like

GPS guidance systems and telematics for enhanced efficiency and precision.

Overall, the operational principles of a tractor revolve around providing power, control, and versatility for performing a wide range of agricultural and industrial tasks efficiently and effectively.

Tractors come in various types, each designed for specific tasks and operating conditions. Here are the different types of tractors commonly used in agriculture and other industries:

1. Utility Tractors: Utility tractors (see Figure 6 and Figure 10) are versatile machines designed for general-purpose tasks on farms, construction sites, and landscaping projects. They typically have horsepower ratings ranging from 20 to 100 and are used for tasks such as hauling, mowing, tilling, and operating implements like loaders and backhoes.

2. Row Crop Tractors: Row crop tractors, such as shown in Figure 11 and Figure 12, are specialized machines used in row-crop farming, where crops are planted in rows with specific spacing between them. These tractors have adjustable wheel widths to navigate between rows without damaging crops. They often feature high ground clearance and narrow profiles to minimize crop damage while providing sufficient power for planting, cultivating, and harvesting row crops like corn, soybeans, and cotton.

3. Orchard Tractors: Orchard tractors are designed for use in orchards and vineyards where narrow rows and low-hanging branches require specialized equipment. These tractors typically have compact dimensions, low profiles, and adjustable wheelbases to manoeuvre easily between rows and under tree canopies. They are used for tasks such as pruning, spraying, and harvesting fruits and nuts.

4. Garden Tractors/Lawn Tractors: Garden tractors are smaller, lighter-duty machines used for landscaping, gardening, and lawn maintenance tasks on residential properties, parks, and golf courses, such as shown in Figure 7. They often feature attachments like lawn mowers, tillers, and snow blowers, making them suitable for a wide range of outdoor maintenance tasks.

5. Compact Tractors: Compact tractors, see Figure 9 for an example, are smaller versions of utility tractors designed for use in confined spaces or on smaller properties. They typically have horsepower ratings ranging from 15 to 40 and are used for tasks such as mowing, landscaping, and light-duty hauling. Compact tractors are popular among hobby farmers, landscapers, and homeowners with smaller properties.

6. Subcompact Tractors: Subcompact tractors, as shown in Figure 8 for an example, are the smallest and lightest tractors available, often used for light-duty tasks on small farms, nurseries, and residential properties. They typically have horsepower ratings below 25 and are designed for tasks such as mowing, tilling, and hauling light loads. Subcompact tractors are easy to manoeuvre and transport, making them ideal for homeowners and hobby farmers with limited storage space.

7. Specialty Tractors: Specialty tractors are designed for specific tasks or industries, such as turf maintenance, airport ground support, and industrial material handling. These tractors often feature specialized attachments, controls, and safety features tailored to their intended applications.

Each type of tractor offers unique features and capabilities suited to different tasks and operating conditions. By selecting the appropriate type of tractor for the job, operators can maximize efficiency, productivity, and safety in agricultural and industrial settings.

Figure 6: John Deere 4520 compact utility tractor.Marcus Qwertyus, CC BY-SA 3.0, via Wikimedia Commons.

In today's market, it's evident that lawn tractors serve a broader purpose beyond mere mowing. It's essential to distinguish between mowers and tractors specifically designed for lawn and garden tasks. Typically, top-tier lawn tractors not only incorporate mowing capabilities but also offer a range of additional functionalities. Featuring horsepower (HP) ratings ranging from 18 to 25, these compact units typically boast wheelbase dimensions of about 45 to 50 inches and are often equipped to accommodate various attachments, including 3-point hitches (Boyce, 2021).

The John Deere E180 serves as an exemplary model within this category. Powered by a robust 25 HP V-Twin extended series engine, this machine delivers forward speeds of up to 5.5 miles per hour and reverse speeds of 3.2 miles per hour. Like many counterparts in its class, the E180 boasts an electric power take-off (PTO) feature. With dimensions measuring 45.7 inches in height and 76.5 inches in length, and a wheelbase of 48.9 inches, this lawn tractor weighs 531 pounds (excluding fuel) (Boyce, 2021).

Figure 7: Example of a lawn tractor. Anon, Public domain, via Wikimedia Commons.

Aligned with the versatile nature of John Deere equipment, the E180 is designed to accommodate a range of attachments. In addition to utility carts and riding covers, options include mulch kits, spreaders, and sprayers. While these tractors are undoubtedly tailored for mowing tasks with decks ranging from 42 inches and above, they also excel in hauling mulch carts, snow blades, and aerators. Pricing for this category of tractors can vary significantly, given the abundance of used models available.

Enthusiasts of the rural lifestyle recognize the significance of a reliable sub-compact tractor, which stands as one of the most sought-after models for both professional producers and hobbyists. Generally characterized by their two- to three-cylinder diesel engines boasting 15 to 25 horsepower (11.2 to 18.6 kilowatts), these machines are equipped with 3-point hitches and are versatile enough to function as lawn tractors for expansive properties. The finest sub-compact tractors offer substantial lift capacities and are often designed to accommodate bucket attachments, as well as rotary mowers for use underneath or behind.

The Kubota BX1880, BX2380, and BX2680 series exemplify these attributes, featuring gross horsepower ranging from 16.6 to 24.8 (12.4 to 18.5 kW), and PTO ratings from 13.7 to 19.5 (10.2 to 14.5 kW). With loader lift capacities reaching 739 pounds (335 kilograms) for a rigid bucket and 613 pounds (278 kilograms) for the QA bucket, these units boast three-point linkage capacities of 680 pounds (308 kilograms), equipped with a Category I 3-point hitch, and are outfitted with four-wheel drive capability. Dimensions vary across models, with lengths ranging around 95.5 inches (242.6 centimetres), widths spanning 44 to 45 inches (111.8 to 114.3 centimetres), and heights of up to 86 inches (218.4 centimetres), while their weights fall between 1,300 and 1,600 pounds (589.7 to 725.7 kilograms), featuring a wheelbase of 55 inches (139.7 centimetres) (Boyce, 2021).

Figure 8: Kubota sub-compact tractor. Mick from Northamptonshire, England, CC BY 2.0 , via Wikimedia Commons.

Sub-compact tractors excel at tasks such as mowing pastures or expansive yards, transporting dirt or mulch, clearing snow, and managing small-scale hay operations. Equipped with the appropriate rotary tiller, they are also well-suited for cultivating large gardens.

While smaller machines might suffice for the rural lifestyle enthusiast, compact tractors are typically required for larger operations. Defined by the American Society of Agricultural Engineers as weighing 4,000 pounds (1,814 kilograms) or less, equipped with a 3-point hitch, and 40 to 60 horsepower (29.8 to 44.7 kilowatts), the best compact tractors can typically lift more than 2,000 pounds (907 kilograms) with a loader. Ideal for mowing, road maintenance, and snow removal, these compact tractors can also till up large gardens and haul small commercial loads (Boyce, 2021).

The Compact Farmall 45C CVT is a member of the CASE IH family of machines and brings a 45 horsepower (33.6 kilowatts) engine and

36 horsepower (26.8 kilowatts) PTO. The tractor is actually built to accommodate an enclosed cab. The tractor measures 10.68 feet (3.25 meters) with a 69.1-inch (175.5-centimetre) width, height of 98.4 inches (249.9 centimetres), and wheelbase of 73.2 inches (185.9 centimetres). The Category I 3-point hitch has a 2,469-pound (1,120-kilogram) lift capacity. The tractor's dry weight is 3,770 pounds (1,710 kilograms) and can handle 2,000 pounds (907 kilograms) with a front-loader.

Figure 9: Australis 40hp compact tractor. ModelTMitch, CC BY-SA 4.0 via Wikimedia Commons.

Compatible with front-end loaders, backhoes, and rotary tillers, the compact tractors are capable of fieldwork as well as grain-hauling and any kind of snow removal or hay production. Great for small farms or businesses involved in landscaping, livestock hauling, or road work. Prices on compact models vary widely based on attachments and region, but figure $30,000 and up unless a great used machine is found (Boyce, 2021).

The utility tractor is loosely defined as having horsepower (HP) between 40 and 100 — sometimes more. Years ago, the compact tractor filled this role, and today there is still considerable crossover between these categories. Several hybrids exist as compact utility tractors, which is easy to understand with the range of sizes being so vast. Most of these models will have 4-cylinder diesel engines and may feature transmissions with gearing such as 8/8, 16/16, or even 32/32. Utility tractors also incorporate significantly more electronics and even Artificial Intelligence (AI) systems these days.

The John Deere 6 Series Utility Tractors offer horsepower between 105 and 250 and can handle anything one wants to throw at them. The 625OR brings a maximum horsepower of 275 with 253 at the PTO ramped to 1,000 RPM. This tractor features a proprietary 8.4-inch CommandCenter display and an infinitely variable transmission system with CommandPRO joystick driving strategy. Capable of reaching 26 miles per hour on the road, the Category 3/3N rear hitch has a maximum capacity of 22,900 pounds. Engine-wise, the diesel features a six-cylinder configuration. The tractor's wheelbase is 114 inches (289.6 centimetres), and it weighs 21,100 pounds (9,574 kilograms) when empty. Built for heavy loads, the maximum permissible Flange axle weight is 33,000 pounds (14,969 kilograms) with 29,652 pounds (13,448 kilograms) (Boyce, 2021).

Figure 10: John Deere 650R. John Deere 650R by Keith Evans, CC BY-SA 2.0 , via Wikimedia Commons.

In technical terms, all tractors fall under the category of agricultural tractors, although another common designation is row crop tractors. Take, for instance, the John Deere 8 Series, which offers a power range of 230 to 410 HP and is available in wheel, two-track, and four-track configurations. Engineered with robust hydraulics to manage large planters, these models epitomize the classic image of tractors often envisioned by urban dwellers.

Row crop tractors represent versatile machinery suitable for various agricultural applications. However, their optimal performance is evident in fields where farmers are planting crops in rows, aligning with their designated purpose.

These tractors boast elevated ground clearance, responsive steering mechanisms, well-suited row-spacing capabilities, compatible attachment implements, robust power lift systems for heavy-duty tasks, and a host of other features tailored to agricultural needs.

Consider the 8RX 410 Four-Track Tractor, which boasts a maximum engine output of 443 HP from its six-cylinder engine. Equipped with Integrated Intelligence, including features like AutoTrac and StarFire receiver, this tractor is fitted with a 10-inch display, specifically the 4600 Generation 4 CommandCenter. Its rated PTO power stands at 310 HP, and it utilizes the e23 PowerShift transmission with Efficiency Manager. The rear hitch is classified as Category 4N/3, offering a standard lift capacity of 20,000 pounds at 24 inches behind the hitch-point (Boyce, 2021).

Figure 11: John Deere 8430. Hanna Zelenko, CC BY-SA 3.0, via Wikimedia Commons.

The base machine weight is 43,300 pounds, with options for tread spacing at 76-, 80-, 88-, or 120-inch intervals. Track width selections include 18-, 24-, or 30-inch variants, while the wheelbase measures 127.36 inches (Boyce, 2021).

Whether it's plowing, harrowing, leveling, weed control, seed drilling, tilling, or cultivating, the row crop tractor excels in handling these agricultural operations effortlessly.

Figure 12: Case IH Magnum 340 with Stoll ProfiLine FZ 100 front-loader. Otto Bolle, CC BY-SA 4.0, via Wikimedia Commons.

Driving and controlling tractors pose greater challenges compared to operating standard cars, necessitating proper training in handling such substantial and potent vehicles before usage. Tractors exhibit several distinctive features, including:

1. Engine Type: Typically powered by diesel engines.

2. Cornering: Demonstrates poor cornering abilities.

3. Driver Protection: Equipped with roll bars for safety.

4. Terrain Usage: Primarily utilized off-road.

5. Brakes: Braking systems may be less effective.

6. Auxiliary Power: Capable of powering other tools or equipment.

7. Centre of Gravity: Exhibits instability with a high centre of gravity.

8. Towing Ability: Can tow heavy implements and equipment.

9. Acceleration: Characterized by slow acceleration.

10. Controls: Features numerous controls to master.

Tractors pose significant risks when operated improperly, largely due to operator errors such as inexperienced handling and lapses in concentration. The complexities of driving tractors necessitate comprehensive training, as they:

- Are substantially heavier than cars, resulting in challenges in stopping, turning, and slow acceleration.
- Have a high centre of gravity, increasing the likelihood of rollovers, hence requiring roll cages or roll over protective structures (ROPS).
- Possess a multitude of levers, switches, and controls, distinct from those found in cars and positioned differently.
- Commonly feature diesel engines that are less prone to stalling, potentially causing sudden forward movement and risking injury to operators or bystanders.
- Primarily operate off-road and are not designed for safe use on paved surfaces, demanding careful manoeuvring on roads.
- Can tow machinery and power other tools or equipment, with power take-offs and three-point linkages presenting inherent

dangers if mishandled.

Prior to operating any tractor, it is essential to thoroughly review its operator's manual. Familiarizing oneself with safety protocols, locating controls, and adhering to all safety warnings specific to the tractor is paramount. When mounting or dismounting, always use the left-hand side of the tractor while it is stationary; attempting this while the tractor is in motion poses significant risks. Before ignition, certain precautions must be taken, including securing the seat belt if a Roll Over Protective Structure (ROPS) is installed, ensuring the tractor is in neutral with the handbrake engaged, disengaging all power take-offs, drives, and hydraulics, and ensuring all bystanders are at a safe distance. Performing maintenance or repairs on a running tractor should be avoided, and all guards and safety features must be maintained intact. Unless explicitly designed for such, passengers should not be transported on the tractor. It is crucial to strictly adhere to the tractor manufacturer's instructions when hitching loads, as improper hitching, especially with loads positioned too high, can lead to rollovers. Extreme caution should be exercised on sloping terrain, favouring driving up and down slopes over traversing across them. Additionally, maintaining a safe distance from overhead powerlines is imperative, with a minimum of 3 meters for lines carrying up to 132 kilovolts and 6 meters for higher voltage lines. Finally, wearing appropriate safety attire suitable for the task at hand is essential.

Identifying potential sources of harm associated with plant equipment, commonly referred to as hazard identification, involves considering both the intrinsic properties of the equipment and the operational context. This includes assessing factors such as electrical, hydraulic, and mechanical components, moving parts, load-bearing capacity, operator safety features, types of loads handled, terrain conditions, and operational practices. Thorough inspections of each piece of plant equipment in use are crucial, along with gathering feedback from workers regarding their experiences with operating, inspecting, or maintaining the equipment. Collaboration with equipment owners, especially when utilizing hired or leased equipment, is essential to identify potential hazards and ensure compliance with safety standards.

Assessing risks involves evaluating the likelihood and potential severity of harm resulting from identified hazards. Consideration should be given to factors such as the severity of potential injuries or illnesses, frequency and duration of exposure to hazards, site-specific conditions, environmental factors, operational practices, competency of operators, and adequacy of training, supervision, and safety protocols. Prompt action should be taken to mitigate identified risks, prioritizing measures based on severity and urgency.

Conducting comprehensive safety protocols and risk assessments is imperative to ensure the safe operation of plant equipment and effectively mitigate potential hazards.

Tractors are frequently encountered pieces of equipment in rural workplaces, yet they are also associated with a notable number of fatalities, particularly in incidents involving rollovers and runovers. These incidents occur regardless of the farm's size or the age and type of the tractor in use.

Agricultural tractors encompass a wide array of vehicles, varying significantly in weight from approximately half a ton to over 25 tonnes. Tractors may exhibit diverse configurations, being either rigid or articulated, equipped with tyres or tracks, featuring two-wheel drive, front-wheel assist, or four-wheel drive, and having single or multiple-wheeled per axle setups.

Addressing safety concerns during the design, manufacture, and operation stages is crucial. This includes implementing features such as roll-over protective structures (ROPS), falling object protective structures (FOPS), guards (e.g., for power take-offs), ensuring safe means of access, mitigating noise and ultraviolet radiation exposure, and incorporating other measures to safeguard operator health and safety, such as seat belts.

Tractors are often equipped to handle various attachments, each introducing its own set of hazards. Thus, careful planning and selection of appropriate equipment for the intended task are essential. It's important to thoroughly assess the tasks at hand, consider the specific equipment required, and evaluate the terrain to ensure the selection of the appropriate tractor-attachment combination.

When utilizing tractor attachments, such as front end loader (FEL) attachments, it's imperative to adhere strictly to their intended purposes. For instance, attempting to lift large round hay bales with an FEL bucket poses significant safety risks.

Transporting passengers on mobile plant, including tractors, is strictly regulated. Passengers must be provided with a level of protection equivalent to that of the operator, typically requiring proper passenger seating and roll-over protection.

A roll-over protective structure (ROPS) is engineered and constructed with the purpose of mitigating the risk of fatalities or injuries to tractor operators in the event of a rollover in any direction. While ROPS does not prevent rollovers, it serves to shield the operator from being crushed if the tractor overturns.

Tractor rollovers can transpire in various settings, including level terrain, uneven surfaces, slight or steep slopes, edges of depressions, contour banks, water courses, and during the towing or pulling of loads, regardless of their weight or stability.

To maximize safety, ROPS should always be utilized in conjunction with a seat belt. This combination prevents the operator from being propelled into the crush zone in the event of an incident. For situations where tractors operate in close proximity to buildings or trees, or indoors, a fold-down ROPS with a locking mechanism may offer greater practicality.

If a tractor is intended for tasks such as tree felling or any other activities that pose a risk of falling objects endangering the operator, it should be equipped with a falling object protective structure (FOPS). FOPS consists of a system comprising structural elements and mesh sheeting affixed to the tractor to shield the operator from potential hazards such as branches, rocks, or bales.

Manufacturers are obligated to ensure that every FOPS meeting technical standards is clearly and permanently labelled with specific information, including:

- The name and address of the FOPS manufacturer
- FOPS identification number
- Make, model, or serial number of the tractor(s) for which the structure is designed
- The relevant Standard or other acceptable technical standard complied with by the structure
- Any additional information deemed pertinent by the manufacturer, such as installation, repair, or replacement instructions.

Tractor access platforms offer a tangible barrier between the tractor operator and its wheels during mounting or dismounting procedures, as depicted in Figure 4. These mechanisms serve to prevent operators from being caught between the front and rear wheels of a tractor, thereby reducing the likelihood of being pulled under the wheel in case of an unexpected movement. Several leading tractor manufacturers integrate safe access platforms into their machine designs, while aftermarket suppliers offer engineered access platforms for older tractor models.

A tractor power take-off (PTO) is a mechanical device that transfers power from the tractor's engine to various types of equipment or implements. It typically consists of a rotating shaft connected to the tractor's engine through a gearbox or transmission system. The PTO shaft extends outwards from the rear or sometimes the front of the tractor and provides rotational power to attached implements such as mowers, tillers, balers, or pumps.

Figure 13: Power take-off (PTO) at the rear end of a John Deere tractor. © User:bdk, CC BY-SA 3.0, via Wikimedia Commons.

The rotational power generated by the tractor's engine is transmitted through the PTO shaft to the connected implement, allowing it to perform its intended function. PTOs come in various sizes and configurations to accommodate different types of equipment and power requirements.

PTOs are widely used in agriculture, construction, and other industries where tractors are employed for powering implements and machinery. They provide a convenient and efficient way to utilize the tractor's engine power for a wide range of tasks, making them essential components of modern tractor systems. Proper safety precautions must be followed when using PTO-driven equipment to prevent accidents and injuries.

Figure 14: PTO shaft is shown connected to the Power Take Off of a farm tractor. Jesster79, CC BY-SA 3.0, via Wikimedia Commons.

Power take-off (PTO) attachments and protective measures: The PTO installed on a tractor can rotate at speeds exceeding 500 RPM, presenting a significant entanglement hazard for operators. Regular inspections of the PTO, shaft, and universal joints should be conducted to identify signs of wear, with all guards promptly maintained or replaced upon detecting damage or wear.

The selection of suitable attachments for specific tasks is paramount, as is the proper fitting of an appropriate PTO guard. Numerous injuries have occurred due to incorrect PTO shaft fitting or inadequate operator training in its usage.

The primary guard, known as the power output coupling (POC) guard, must be permanently affixed to the tractor, with provisions for mobility while ensuring secure positioning during tractor operation. Likewise, the implement power input coupling (PIC) guard should be permanently attached to the implement, movable if necessary, and securely held in place during use, eliminating any potential 'nipping points' where body parts or clothing may become caught.

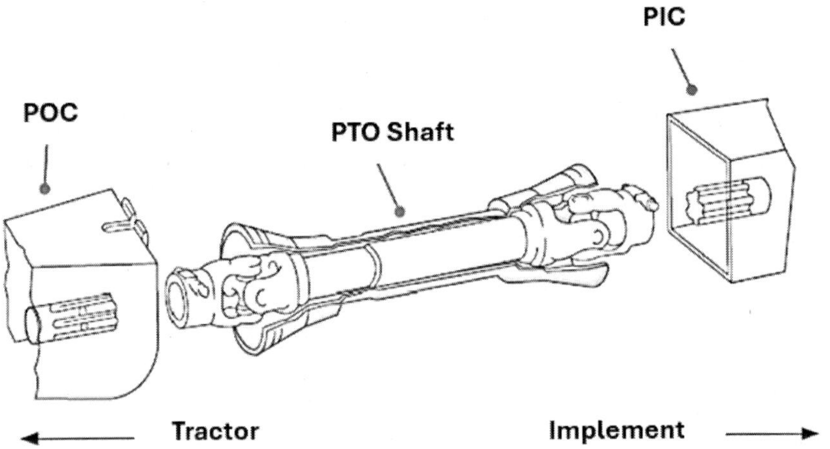

Figure 15: Tractor PTO guard.

The PTO shaft guard should extend sufficiently into both the tractor's POC guard area and the implement's PIC guard area, offering maximum practical coverage. Guards may be of rotating or non-rotating types, with non-rotating guards requiring a means of restraint to prevent unintended movement. Prior to purchasing guards, ensure they comply with relevant Standards and are of appropriate size and length for the drive shaft. If uncertain, consult with the guard manufacturer or supplier, taking into consideration the shaft's vertical and lateral movements during operation.

Regarding all guards:

- Regularly inspect all guards for signs of wear and damage, such as daily checks during usage. Any guards found to be damaged must be replaced before further use.

- When protection is required in the PTO drive line, torque limiters, free wheels, or clutches should be positioned at the power input connection (implement end) of the PTO drive shaft.

- The devices used to prevent the rotation of the shaft guard should not serve as a means of supporting the PTO drive shaft or guard when the machine is disconnected.

- When the machine is idle, support the drive shaft and guard on the provided cradle. If no cradle is available, use alternative means to support the shaft and guards to ensure equivalent protection against damage.

When operating PTO equipment:

- Disengage the power drive.

- Shut off the tractor engine.

- Confirm that controls are in neutral and apply the handbrake.

- Remove the engine key.

- Wait for all movement to cease before attempting to clear any blockage, and utilize a tool for this purpose.

Front end loader (FEL) attachments mounted on tractors are extensively utilized in the rural sector and encompass various types, such as:

- Single or multi-purpose buckets

- Pallet forks

- Bale and silage spikes

- Bale and silage clamps and grapples

- Blades and scrapers

- Lifting jibs

Typically, an FEL attachment connects to a tractor via a sub-frame that is permanently affixed to the tractor, as depicted in Figure 6. A quick detach and locking system from the sub-frame typically ensures positive engagement and secure retention of the FEL attachment under diverse operating conditions.

Figure 16: Massey Ferguson front loader tractor. Acabashi, CC BY-SA 4.0 , via Wikimedia Commons.

Front end loader (FEL) attachments should come equipped with a support stand positioned to elevate the arms to the appropriate height, facilitating the tractor's ingress or egress when connecting or disconnecting the arms. This support stand ought to be situated on a stable, level surface capable of bearing the weight of the unhitched FEL attachment.

Before utilizing a tractor with an FEL attachment, a comprehensive risk assessment must be conducted, considering various factors including operator proficiency, machine specifications, and environmental conditions. Factors to assess regarding the use of an FEL attachment include:

- Operator's competency level and familiarity with the equipment
- Potential for objects or loads to shift or fall onto the operator
- Front axle, wheels, and tyre capacity to withstand the weight of a fully loaded FEL attachment
- Hydraulic system's lifting capability
- Clearance between tractor front tyres and FEL attachment frame to prevent contact during turns
- Stability of the tractor when operating with a loaded FEL attachment
- Suitability of the selected FEL attachment for the lifting task
- Operating conditions such as material density, load dimensions, speed, load height, and terrain surface characteristics
- Possibility of exceeding the Rated Operating Load (ROL) of the FEL attachment, leading to rear axle instability and loss of traction

Installation of an FEL attachment on a tractor should not proceed unless the tractor is equipped with a Roll Over Protective Structure (ROPS) or a cabin featuring ROPS. Ideally, the ROPS should be a four-post structure or positioned forward of the operator to provide protection against objects rolling back from the bucket or lifting mechanism. Measures to prevent load rollback include the use of specialized lifting attachments, a level lift system, a rollback guard, and a lifting height limiting device. When there is a risk of objects or materials falling onto the operator, the ROPS should be fitted with a Falling Object Protective Structure (FOPS).

Each FEL attachment should be matched to the tractor, with a decal or plate specifying its ROL for the particular tractor model. Guidance should be sought from the tractor manufacturer to ensure compatibility with the engine capacity and hydraulic system for satisfactory performance. Additionally, rear weights or ballast may be added to improve stability by shifting the centre of gravity rearward. Following the tractor manufacturer's instructions and recommendations is crucial when adding ballast.

Quick-release hydraulic couplings facilitate easy attachment and detachment of FEL attachments, with clear markings to avoid incorrect connections. Hydraulic pressure should be released before disconnecting to ensure safe operation.

The main components of a tractor serve distinct functions essential for its operation. The engine, functioning as the primary power source, drives various tractor activities by transmitting power to both the wheels and the Power Take-Off (PTO) through transmission components. This critical role establishes the engine as the central component, akin to the tractor's heart, typically employing four-stroke internal combustion engines.

Figure 17: Tractor components. Back Image - Fruitman cz, CC BY-SA 4.0, via Wikimedia Commons.

The powertrain, comprised of the clutch, gearbox, differential, and final reduction, is equally vital, serving as the transmission mechanism that ensures power from the engine reaches the tractor's wheels effectively. The clutch facilitates initial engine movement, gear shifting, and synchronization between engine and gearbox, establishing a vital link in the power transfer process.

Integral to the powertrain, the gearbox facilitates the transmission of engine power from the clutch to the wheels at desired revolutions and speeds. Meanwhile, the differential enables safer cornering by allowing outer wheels to rotate more than inner wheels, enhancing driving stability during turns.

The PTO, or Power Take-Off, is instrumental in transmitting motion to rotating machinery, facilitating various field tasks. Available in configurations operating at either 540 or 1000 revolutions per minute, the PTO enables the operation of tools dependent on rotational motion.

The steering system dictates the tractor's directional movement, providing steering control over the front wheels or track systems. Different steering types, including axle rotating and articulated systems, offer diverse steering functionalities tailored to specific tractor designs and operational requirements.

Crucial for driving safety, the brake system enables speed modulation and halts tractor motion when necessary. Consisting of both parking and foot brakes, this system ensures operational control and safety during various tractor manoeuvres.

Furthermore, the rear linkage arrangements on tractors facilitate the connection of tools and machinery for towing and implement usage. Rear hitches provide towing capabilities, offering versatile towing layouts such as high drawbar connection, oscillating drawbar, perforated drawbar, and quick-release tow hook, catering to diverse tractor usage scenarios and objectives.

The centre of gravity (COG) of a tractor refers to the point within the tractor where its weight is evenly distributed in all directions. It's a crucial concept in tractor design and operation because it affects stability and balance. The COG is typically located somewhere near the geometric centre of the tractor but can vary depending on factors such as the distribution of components, attachments, and the load being carried.

Understanding the tractor's centre of gravity is essential for safe operation, especially when navigating uneven terrain or carrying heavy loads. Tractors with a lower centre of gravity are generally more stable and less prone to tipping over, whereas those with a higher COG may be more susceptible to tipping, particularly when turning sharply or operating on sloped surfaces.

Farmers and operators need to be aware of the tractor's COG to prevent accidents and ensure safe operation. Proper loading, distribution of weight, and adherence to manufacturer recommendations regarding maximum load capacities are essential for maintaining stability and minimizing the risk of rollovers or other accidents.

A tractor becomes unstable and may tip over when its centre of gravity shifts beyond the "stability baseline," which is represented by an imaginary line connecting all the wheels, as shown in Figure 18.

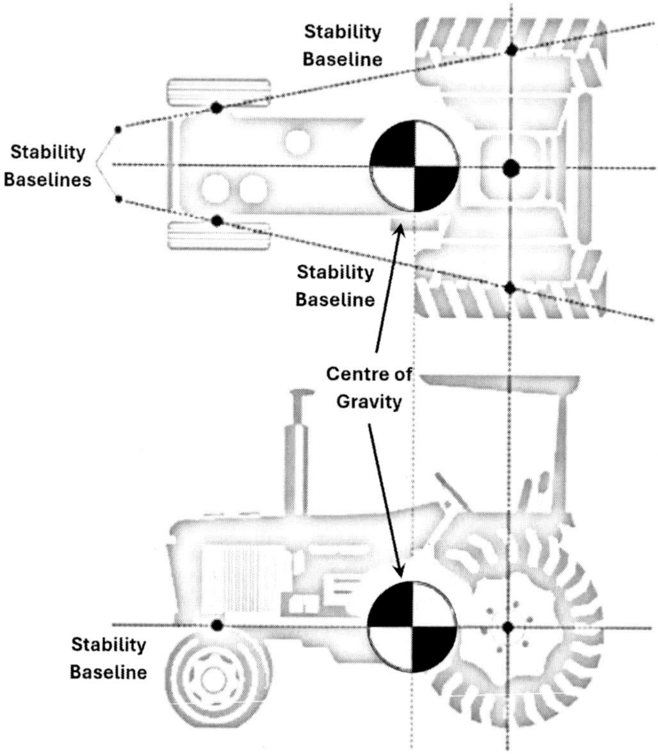

Figure 18: Stability baselines and Centre of Gravity.

Side-mounted implements are attachments or tools that are mounted on the side of a tractor rather than directly behind it. When such implements are attached to a tractor, they shift the centre of gravity of the tractor towards the side where the implement is mounted, as shown in Figure 19. This happens because the weight of the implement adds to the overall mass of the tractor, causing a redistribution of weight and altering the balance of the tractor. As a result, the tractor's stability may be affected, particularly when navigating uneven terrain or making sharp turns. It's important for tractor operators to be aware of this shift in the centre of gravity and adjust their driving accordingly to maintain safe operation.

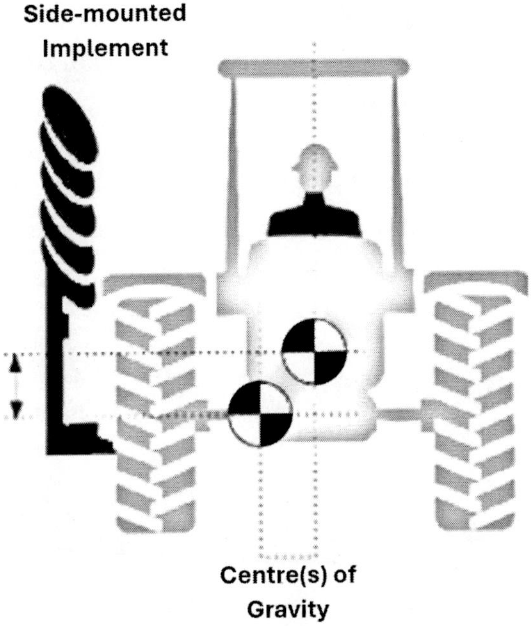

Figure 19: Shift in Centre of Gravity die to side-mounted implement.

Operating the tractor with the mounted implement on the uphill side of the slope means positioning the implement on the side of the tractor that is facing uphill when working on sloped terrain. This practice is recommended for safety and efficiency reasons:

1. Improved Stability: Placing the implement on the uphill side helps to counterbalance the gravitational force pulling the tractor downhill. This positioning can enhance stability by preventing the tractor from tipping over sideways.

2. Better Traction: With the weight of the implement positioned uphill, the tractor's drive wheels have more traction on the

ground. This improves traction and reduces the risk of the tractor slipping or losing control on steep slopes.

3. Easier Manoeuvring: Operating with the implement on the uphill side can make it easier to control the tractor, especially when navigating turns or changes in terrain elevation. It helps to maintain better control over the direction of travel and reduces the likelihood of the tractor veering off course.

Overall, positioning the implement on the uphill side of the slope is a safety measure that helps to mitigate the risks associated with operating tractors on uneven or sloped terrain, promoting safer and more efficient agricultural practices.

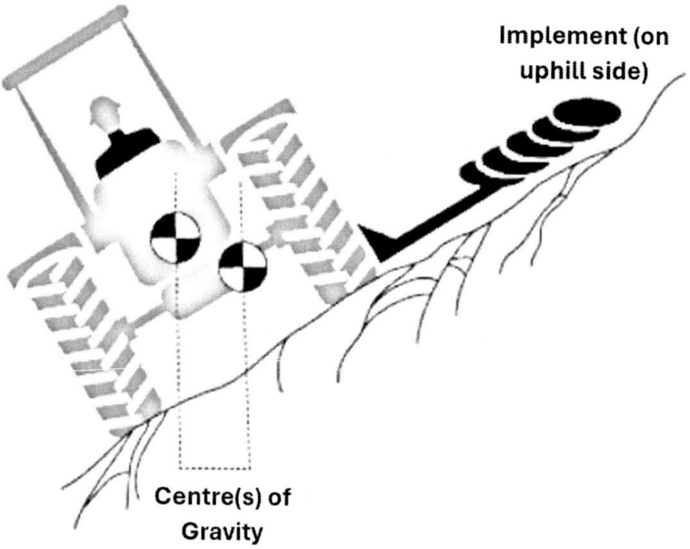

Figure 20: Keeping implement on uphill side to counterbalance gravitational force.

Tractor stability can be enhanced by adjusting its weight distribution using wheel weights and counterweights. Wheel weights are added to the tractor's wheels to increase traction and stability, especially when carrying heavy loads or operating on uneven terrain. Counterweights are typically attached to the rear of the tractor to balance the weight of front-mounted implements or loads. By strategically adjusting the placement and amount of these weights, operators can optimize the tractor's stability, reducing the risk of tipping or losing control, particularly during demanding tasks or on slopes.

Operators should remain vigilant for ground depressions, such as ruts, furrows, or holes, as well as obstructions like rocks and stumps, while operating the tractor. These hazards pose risks to tractor stability, particularly when traversing slopes or traveling at higher speeds. Ground depressions can cause sudden shifts in weight distribution, leading to loss of traction or stability. Similarly, striking obstructions like rocks or stumps can destabilize the tractor, potentially causing it to tip over or veer off course. Operators should exercise caution, adjust speed accordingly, and manoeuvre carefully to avoid these hazards and maintain safe tractor operation.

Instability can cause a tractor to tilt sideways, with its outside wheels acting as pivot points. This sideways tipping force fluctuates depending on the tractor's speed and the sharpness of its turns, with sharper turns increasing the force exerted. Particularly when turning uphill, a tractor is more prone to instability, making it advisable to avoid such manoeuvres whenever feasible or to execute them at a reduced speed. The risk intensifies as the tractor's centre of gravity is elevated, such as when a front-end loader is raised or when spray tanks are mounted high on the tractor chassis.

There are many tractor attachments on the market, including:

1. Front Blade: A front blade, also known as a front-mounted blade or snow blade, is a large blade attachment mounted on the front of a tractor. It is primarily used for pushing and moving materials such as snow, dirt, or debris. Front blades are commonly used for snow removal, landscaping, and earthmoving tasks.

2. Forklift: A forklift attachment allows a tractor to lift and trans-

port heavy pallets, crates, or other materials. It typically consists of two or more hydraulic-powered forks that can be raised and lowered to lift loads. Forklift attachments are commonly used in warehouses, construction sites, and agricultural operations for loading and unloading tasks.

3. Slasher: A slasher, also known as a brush cutter or rotary slasher, is a cutting attachment used for clearing dense vegetation, bushes, and small trees. It typically consists of a rotating blade or blades attached to a spinning shaft, powered by the tractor's PTO. Slashers are commonly used for land clearing, vegetation management, and maintaining pastures.

4. Mower: A mower attachment is used for cutting grass, weeds, and other vegetation. Mowers come in various types, including rotary mowers, flail mowers, and sickle bar mowers, each suited for different terrain and vegetation types. They are commonly used for lawn maintenance, roadside mowing, and pasture management.

5. Auger: An auger attachment is used for drilling holes in the ground for fence posts, planting trees, or installing utility poles. It typically consists of a rotating spiral-shaped drill bit attached to a shaft powered by the tractor's PTO. Augers come in various sizes and configurations to accommodate different hole diameters and depths.

6. Drag Broom: A drag broom attachment is used for sweeping and leveling surfaces such as driveways, roads, and parking lots. It typically consists of a series of stiff bristles or brushes attached to a rotating drum, which is dragged behind the tractor. Drag brooms are commonly used for dust suppression, snow removal, and surface maintenance.

7. Power Broom: Similar to a drag broom, a power broom attachment is used for sweeping and cleaning surfaces. However, unlike a drag broom, a power broom is self-powered and may include additional features such as adjustable brush angles and hydraulic controls for enhanced versatility and efficiency.

8. Loading Platform: A loading platform attachment provides a flat surface for loading and transporting materials, equipment, or livestock. It typically consists of a sturdy platform mounted on the tractor's rear or front end, with sides or rails to contain the load. Loading platforms are commonly used in agricultural, construction, and transportation applications.

9. Rotary Hoe: A rotary hoe attachment is used for tilling and cultivating soil in preparation for planting. It typically consists of multiple rotating tines or blades that break up and aerate the soil. Rotary hoes are commonly used in gardening, landscaping, and agriculture to improve soil structure and promote plant growth.

10. Concrete Mixer: A concrete mixer attachment is used for mixing and pouring concrete on construction sites or for DIY projects. It typically consists of a rotating drum mounted on the tractor's rear or front end, with a hydraulic or mechanical drive system. Concrete mixers are commonly used in construction, roadwork, and building projects.

11. Spraying Equipment: Spraying equipment attachments are used for applying pesticides, herbicides, fertilizers, or other chemicals to crops or vegetation. They typically consist of a tank for holding the liquid, a pump for spraying, and a nozzle or boom for distributing the spray. Spraying equipment attachments are commonly used in agricultural operations for crop protection

and weed control.

12. Disc Plow: A disc plough attachment is used for plowing and turning over soil in preparation for planting. It typically consists of multiple disc blades mounted on a rotating shaft, powered by the tractor's PTO. Disc ploughs are commonly used in agriculture for primary tillage and soil preparation tasks.

Tractors are engineered to couple with ploughs and various implements through a three-point linkage system. This setup involves three connection points, forming a triangular configuration, allowing the implement's weight to be supported by the tractor itself. This arrangement is devised to shift a portion of the implement's weight to the tractor below its centre of gravity, thereby enhancing traction.

Figure 21: John Deere 3050, rear view showing three-point linkage, PTO and swinging drawbar. Peter Mooney, CC BY-SA 2.0 , via Wikimedia Commons.

In contrast, trailers possess wheels and independently bear their weight.

Certain tractors are equipped with 'quick hitches' to affix trailed equipment to the three-point linkage. This repositions the towing point rearward, potentially causing the tractor to exhibit unpredictable behaviour during braking and cornering manoeuvres.

Improperly connecting implements or placing them above the tractor's centre of gravity can lead to a backward rollover of the tractor.

Pulling objects by fastening chains or cables to an elevated point rather than the drawbar can also result in a backward rollover of the tractor.

Using an incorrect hitching pin poses a sudden failure risk, potentially causing the detachment of the towed load.

Additionally, there is a risk of being crushed or run over while attaching an implement to the tractor.

Always adhere to the manufacturer's specifications when attaching implements to the tractor, ensuring that the load is securely affixed well below the tractor's centre of gravity.

Maintain a horizontal and low pull angle whenever possible.

Ensure proper matching of equipment with the appropriate-sized tractor by consulting the manufacturer's recommendations, and utilize counterweights if necessary.

During the process of reversing to connect the implement, ensure that no individual stands between the implement and the tractor. In instances where it's necessary to move the tractor while attaching the implement, helpers should vacate the area between the tractor and implement. If it becomes necessary to incrementally adjust the tractor's position while someone inserts connecting pins, ensure that the tractor doesn't reverse too far initially. Instead, have the helper approach, then gradually inch the tractor forward until the pins can be inserted.

Follow the correct sequence for attaching implements to the three-point linkage: left, right, then centre.

Figure 22: Rear three-point linkage/hitch of a John Deere tractor: 1. External linkage control, 2. Lift rods, 3. Top link (centre arm), 4. Hitch lifting arms, 5. Power take-off (PTO), 6. Drawbar, 7. Hydraulic connections. Sam Craig, CC BY-SA 2.0, via Wikimedia Commons.

The three-point linkage system on a tractor is a mechanism designed to connect and control implements such as plows, mowers, and cultivators. It consists of three connection points arranged in a triangular configuration: two lower lift arms and an upper link. Here's an explanation of the connection methods for using a three-point linkage on a tractor:

1. Lower Lift Arms: These are typically hydraulic arms located on either side of the rear of the tractor. They are used to lift and lower the implement and provide support for the weight of the attached equipment. The lower lift arms are adjustable in length to accommodate different implements.

2. Top Link: The top link is an adjustable bar or rod located above

the lower lift arms. It connects the top of the implement to the tractor's rear hitch point. The top link can be extended or retracted to adjust the angle of the implement relative to the ground.

To connect an implement using the three-point linkage system:

1. Position the Tractor: Park the tractor on a flat and stable surface, ensuring it is safely secured.

2. Align the Implement: Position the implement behind the tractor, aligning it with the lower lift arms and hitch point.

3. Lower the Lift Arms: Lower the lower lift arms using the tractor's hydraulic controls until they are aligned with the implement's attachment points.

4. Attach the Lower Lift Arms: Connect the lower lift arms to the implement's mounting brackets or pins. Ensure they are securely attached and locked in place.

5. Adjust the Top Link: Adjust the length of the top link to match the height and angle of the implement. Connect one end of the top link to the tractor's hitch point and the other end to the implement.

6. Secure the Implement: Double-check all connections to ensure they are secure and properly aligned. Test the attachment by raising and lowering the implement using the tractor's hydraulic controls.

7. Check for Clearance: Verify that there is sufficient clearance between the implement and the tractor when raised and lowered. Make any necessary adjustments to ensure proper clearance.

8. Operate Safely: Once the implement is securely attached, oper-

ate the tractor and implement safely, following all recommended procedures and safety guidelines.

The essential components of the three-point hitch include a pair of lower arms, a top link, and adjustable tie rods and chains that facilitate movement via hydraulic lift and prevent lateral sway. The lower arms, sturdy steel bars with a rectangular cross-section, are hinged at one end to the tractor frame, enabling convenient rotation within the designated operating arc. Similarly, the top link, acting as a robust tie-rod, is also hinged to the tractor body and comprises two oppositely threaded screws with eyelets at one end, screwing into a central female sleeve at the opposite end. By rotating the sleeve, the top link can be lengthened or shortened, allowing for convenient adjustment of the attached equipment's vertical alignment.

The ends of the lower arms and the top link form the "attachment triangle," which must align precisely with the pins of the corresponding triangular linkage frame on the implement. Geometrically precise attachment points ensure proper fit, with standardized dimensions classified into categories for regulatory compliance. These categories, numbered 1 through 4, indicate increasing sizes, with a special sub-category, 1N, designed for narrow-track tractors, characterized by an S-shape configuration to accommodate reduced distance to hinged attachments.

While traditionally operated mechanically, modern three-point hitches can utilize hydraulic cylinders for adjustments, particularly in lifting functions. Hydraulic alternatives offer enhanced control, with external hydraulic cylinders replacing mechanical tie rods in some configurations, providing improved implement handling and safety.

Furthermore, hydraulic enhancements such as the hydraulic third link with an integrated shock absorber and hydraulic three-point hitch stabilizer contribute to smoother operations and increased safety. These innovations mitigate load peaks, control lateral oscillations, and facilitate transitions between floating and rigid positions, ensuring optimal performance and stability during agricultural tasks.

The process of hitching and unhitching implements typically involves coordination between two operators: one driving the tractor and the other on the ground securing the attachment pins with safety pins as the tractor is positioned to align with the attachment points.

This procedure poses significant risks for the ground operator, who must stand between the tractor and the equipment. Accidents, including serious ones, can occur if the tractor's reverse speed is not carefully controlled, potentially leading to crushing injuries between the two machines. Consequently, efforts have been made to enable remote hitching and unhitching, allowing for driver intervention only and eliminating the need for a ground operator.

Typically, attachment triangle pins fit into rotules located at the end of the lower arms and the third point. These rotules can rotate within limited angles to facilitate coupling with the pins. Alternatively, rotules can be equipped with robust pressed steel hooks, designed to withstand high tensile, compressive, and lateral forces. These hooks feature a ratchet mechanism controlled remotely by cables operated by the driver from the driver's seat. Internal mechanisms provide safety locking, supplemented by centring guides and a kick-down function for additional security.

The front linkage is an increasingly favoured attachment aimed at enhancing tractor versatility. It enables combined work, often in conjunction with a power take-off and/or hydraulic sockets, or facilitates the mounting of monolithic ballast. To optimize tractor dimensions, the lower arms of this linkage may consist of two sections, with the outer section capable of folding back at least 90° when not in use. Similarly, the third link is easily removable. However, the hydraulic lift operating this linkage typically offers limited operating alternatives compared to the rear linkage, often limited to position-controlled and floating modes.

To monitor loads, modern tractor models are equipped with electronic lifts utilizing sensors, typically strain gauges, affixed to specific components of the three-point linkage, such as the lower arms or their hinge pins. These sensors detect microscopic deformations generated by dynamic stress on the components, converting them into small changes in electric current. Strain gauges measure deformations and compressions and can efficiently amplify these variations for precise load monitoring. The working principle of strain gauges relies on the variation of their resistive value under mechanical stress, with the support base and adhesive binder ensuring accurate transmission of strain while providing electrical isolation.

Planning and Preparing for Tractor Operations

Planning and preparing for tractor operations involves several crucial steps to ensure the efficiency, safety, and success of the tasks at hand. These include:

1. Assess the Task Requirements: Begin by understanding the objectives and requirements of the tasks to be performed with the tractor. This includes determining the specific activities, equipment needed, and any potential hazards or challenges associated with the job.

2. Review Documentation and Procedures: Access tractor operations manuals, safety guidelines, and any relevant documentation provided by the manufacturer or employer. Review procedures for operating the tractor, maintenance protocols, and safety precautions to ensure compliance with industry standards and regulations.

3. Evaluate Environmental Factors: Consider the environmental conditions in which the tractor operations will take place. Assess factors such as terrain, weather conditions, visibility, and any potential hazards such as uneven ground, obstacles, or overhead obstacles like power lines.

4. Identify Hazards and Risks: Conduct a thorough hazard assessment to identify potential risks associated with tractor operations. This includes recognizing hazards such as rollover risks, entanglement hazards, and environmental dangers. Assess the likelihood and severity of these risks to determine appropriate control measures.

5. Develop a Work Plan: Create a detailed work plan outlining the sequence of tasks, equipment needed, safety measures, and contingency plans. Ensure that the plan accounts for all aspects of the operation, including pre-operational checks, task execution, and post-operational procedures.

6. Select and Prepare Equipment: Choose the appropriate tractor and attachments for the planned tasks based on their specifications and compatibility with the job requirements. Inspect the tractor and attachments for any signs of damage, wear, or malfunction, and perform any necessary maintenance or repairs.

7. Check Personal Protective Equipment (PPE): Identify the PPE required for tractor operations based on the hazards present, such as gloves, safety boots, helmets, or hearing protection. Ensure that all personnel involved in the operation have the necessary PPE and that it is in good condition.

8. Communicate and Coordinate: Coordinate with supervisors, colleagues, or other personnel involved in the operation to en-

sure everyone is aware of their roles and responsibilities. Communicate the work plan, safety procedures, and any specific instructions or precautions to be followed during the operation.

9. Prepare the Work Area: Clear the work area of any obstacles, debris, or hazards that could interfere with tractor operations. Ensure that signage, barriers, and other safety measures are in place to alert others to the ongoing work and restrict access to the area if necessary.

10. Emergency Preparedness: Familiarize yourself and your team with emergency procedures for responding to accidents, injuries, or equipment malfunctions. Ensure that emergency contact information, first aid supplies, and firefighting equipment are readily available and accessible.

Accessing tractor operations documentation involves obtaining manuals, guides, and any relevant instructional materials provided by the tractor manufacturer or employer. Interpretation of this documentation entails understanding operating procedures, safety guidelines, maintenance schedules, and troubleshooting protocols. Applying this information involves implementing the recommended practices and procedures while operating the tractor, ensuring safe and efficient usage.

Obtaining work instructions involves receiving directives from supervisors or team leaders regarding specific tasks to be performed with the tractor. Interpretation requires understanding the objectives, methods, and safety considerations associated with the assigned tasks. Clarifying involves seeking clarification from supervisors or colleagues if any aspects of the instructions are unclear. Confirming involves verifying understanding and receiving approval or confirmation from supervisors before proceeding with the task.

Identifying hazards involves recognizing potential dangers associated with tractor operations, such as terrain conditions, equipment malfunctions, or nearby obstacles. Assessing risks involves evaluating the likelihood and severity of potential accidents or incidents arising from identified hazards. Implementing control measures involves taking proactive steps to mitigate or eliminate risks, adhering to workplace policies and safety regulations, which may include using personal protective equipment, modifying work practices, or implementing engineering controls.

Selecting personal protective equipment (PPE) involves identifying the appropriate gear based on the specific hazards present during tractor operations, such as gloves, safety boots, helmets, or hearing protection. Wearing PPE is essential to mitigate risks and ensure personal safety while operating the tractor or working in its vicinity.

Identifying traffic management signage requirements involves recognizing the need for signage to regulate traffic flow or warn of potential hazards in areas where tractors are operating. Obtaining signage involves acquiring the necessary signs, cones, or barriers from designated storage areas. Implementing signage involves placing them strategically to alert and guide vehicle operators, pedestrians, and other workers in compliance with standard operating procedures and safe work practices.

Selecting tractor equipment and attachments involves determining the appropriate implements needed to perform specific tasks effectively and efficiently. This may include plows, mowers, loaders, or other specialized tools. Proper selection ensures compatibility with the tractor and suitability for the intended task, maximizing productivity and safety.

Obtaining emergency procedures involves familiarizing oneself with protocols for responding to fires, accidents, or other emergencies while operating the tractor. Interpretation involves understanding the steps to take in various emergency scenarios, such as equipment malfunctions, fires, or injuries. Being prepared involves maintaining awareness of emergency exits, first aid supplies, fire extinguishers, and other safety equipment, as well as knowing how to activate emergency response protocols and seek assistance if needed.

Coordinating planned activities involves communicating with supervisors, colleagues, or other personnel to ensure alignment and cooperation in executing tasks involving tractor operations. This may include discussing task objectives, assigning responsibilities, and establishing timelines. Effective communication fosters teamwork, minimizes misunderstandings, and enhances overall safety and productivity on the work site.

The only secure position for individuals on a tractor is within the driver's seat. Any additional riders risk serious injury from unexpected jolts or sharp turns, potentially leading to ejection from the tractor. Therefore, it is strongly advised against requesting rides on tractors or farm equipment.

Even when a tractor is not operational, jumping off can pose significant risks. Factors such as muddy footwear or loose clothing increase the likelihood of mishaps during the descent, potentially resulting in injury.

Most tractors, regardless of age, feature a protruding Power Take-Off (PTO) shaft at the rear. This shaft transfers power to auxiliary machinery like manure spreaders, rotating at speeds ranging from 9 to 15 rotations per second. Tractors are typically equipped with a protective shield covering the PTO shaft. However, if this safety shield is absent, individuals near the rear of the tractor are exposed to considerable danger.

When the rear wheels of a tractor are immobilized, causing the front end to rise, the tractor can overturn in less than a second. Common factors contributing to rearward tractor turnovers include being stuck in mud or snow, obstruction of wheel movement by chains or boards, ascending excessively steep hills, releasing the clutch abruptly in low gear at high engine speeds, or hitching a load above the drawbar.

Sideways tractor rollovers are another potential hazard, occurring when the tractor is driven on overly steep hillsides, the front-end loader is elevated too high for the load or turning on a hill, or when the tractor ventures too close to the edge of steep slopes like roadside ditches.

Tractor operations pose various hazards, including:

1. Rollovers: Tractors are prone to rollovers, especially when operated on uneven terrain or slopes. Rollovers can result in serious injuries or fatalities to the operator.

2. Entanglement: Tractors have moving parts such as power take-off (PTO) shafts, belts, and rotating components that can cause entanglement with loose clothing, hair, or body parts, leading to severe injuries or even amputations.

3. Falls: Operators can fall from tractors while mounting or dismounting, particularly if they do not maintain three points of contact or if the tractor is in motion.

4. Struck-by incidents: Tractors may collide with stationary objects such as trees, buildings, or other vehicles, leading to injuries to operators or bystanders.

5. Crush injuries: Tractors can crush operators or bystanders if they overturn or if heavy loads are mishandled.

6. Overexertion: Operating a tractor for extended periods can lead to fatigue and overexertion, increasing the risk of accidents due to reduced alertness and coordination.

7. Noise exposure: Tractors generate high levels of noise, which can lead to hearing loss if operators are not adequately protected with ear protection.

8. Chemical exposure: Exposure to pesticides, fertilizers, or other

chemicals used in agricultural operations can pose health risks to tractor operators if proper precautions are not taken.

9. Visibility issues: Tractors may have blind spots, obstructed views, or limited visibility, increasing the risk of collisions with other vehicles or hazards.

10. Mechanical failures: Malfunctions or failures of tractor components such as brakes, steering, or hydraulics can lead to accidents if not properly maintained or addressed promptly.

Identifying hazard points includes:

Pinch Points: Pinch points, also known as mangled or maimed points, occur when two rotating objects converge, with at least one moving in a circular motion.

Wrap Points: Wrap points refer to rotating shafts where items like cuffs, sleeves, pant legs, long hair, or even loose threads can become entangled. It's essential to inspect all equipment for potential wrap points and, whenever feasible, install shields to mitigate risks.

Shear Points: Shear points are formed when the edges of two objects come into close proximity, capable of cutting through soft materials.

Crush Points: Crush points emerge when two objects move towards each other or when a stationary object is approached by a moving one. An example of this is hitching tractors to implements, which can create a crush point.

Stored Energy Hazards: Hazards related to stored energy are prevalent in pressurized systems like hydraulics, compressed air, and springs. The sudden release or application of pressure in these systems can lead to crushing accidents.

Pull-in Points: Pull-in points are mechanisms designed to draw in crops or other materials for processing. Examples include combine headers, windrow pickups, and grinders. Identifying and understanding these points is crucial for implementing defensive farming practices.

A tractor pre-start check is a routine inspection conducted before operating the tractor. Its purpose is to ensure that the tractor is in proper working condition, identify any potential issues or hazards, and mitigate risks to the operator, bystanders, and property. Performing a pre-start check is essential for safety, as it helps prevent accidents, breakdowns, and damage to equipment.

A tractor pre-start check should be performed before each use or shift, ideally at the beginning of the workday or whenever the tractor is about to be operated. By conducting the pre-start check before each use, tractor operators can ensure that the equipment is in optimal condition and safe to operate.

Performing the pre-start check before each use allows operators to identify any potential issues or hazards that may have arisen since the last operation. This proactive approach helps prevent accidents, breakdowns, and damage to equipment by addressing any problems before they escalate.

Additionally, conducting the pre-start check at the beginning of the workday sets a routine for safety and maintenance, emphasizing the importance of thorough inspections before operating the tractor. It also helps establish a culture of safety among tractor operators, encouraging them to take responsibility for the condition and maintenance of the equipment they operate.

Ultimately, performing the pre-start check before each use ensures that the tractor is in proper working condition, mitigating risks to the operator, bystanders, and property while maximizing safety and productivity.

The pre-start check should include the following components:

1. Visual Inspection: Examine the exterior of the tractor for any signs of damage, leaks, or loose components. Check for fluid leaks, such as oil, hydraulic fluid, or coolant, which could indicate potential issues.

2. Tyres: Inspect the tyres for proper inflation, tread wear, and damage. Ensure all tyres are in good condition and free from punctures or cuts that could affect stability or traction.

3. Fluid Levels: Check the levels of engine oil, hydraulic fluid, coolant, and fuel. Topping up fluids as necessary ensures proper lubrication, cooling, and operation of the tractor.

4. Battery: Verify that the battery terminals are clean, tight, and free from corrosion. Check the battery voltage to ensure it has sufficient charge for starting the engine.

5. Lights and Signals: Test all lights, including headlights, taillights, brake lights, turn signals, and hazard lights, to ensure they are functional. Proper lighting is crucial for visibility, especially when operating the tractor at night or in low-light conditions.

6. Controls and Instruments: Check that all controls, switches, and gauges are functioning correctly. Test the throttle, brakes, steering, and hydraulic controls to ensure they respond as expected.

7. Safety Devices: Inspect safety features such as seat belts, roll-over protection structures (ROPS), and guards for power take-off (PTO) shafts and other rotating parts. Ensure all safety devices are in place and operational.

8. Attachments: If using attachments, inspect them for damage, wear, and proper installation. Ensure all attachments are securely mounted and free from defects that could affect performance or safety.

9. Documentation: Verify that all required documentation, such as the operator's manual, safety guidelines, and maintenance records, is present and accessible.

By performing a thorough pre-start check, tractor operators can identify and address potential issues before they escalate into safety hazards or equipment failures. This proactive approach helps ensure the safe and efficient operation of the tractor while minimizing the risk of accidents, downtime, and costly repairs.

Handling hay bales or other heavy objects can elevate the risk of rollover, while hay and wrapped silage might slide back down the loader's arms onto the operator if the load is elevated too high or fall from a stack onto the tractor driver.

To mitigate this risk:

- Utilize the appropriate attachment for the task and adhere to the manufacturer's guidelines.
- Employ a hay spike for handling hay and a grab for wrapped silage, avoiding the use of flat forks or buckets.
- Transport bales slowly and maintain them as close to the ground as possible.
- Verify the tractor's lift capacity and utilize a counterbalance if necessary.
- Refrain from carrying bales stacked higher than the back frame of the forks; consider extending the frame height safely for accommodating additional bales.
- Exercise caution to prevent the implement from tilting upwards, thereby preventing bales from slipping down the lift arms.

Operating a Tractor

When ascending or descending from a tractor, ensure consistent contact with three points, such as two hands and one foot, either on the tractor or the ground. Depart facing towards the tractor, maintaining the same approach used during embarkation. Avoid leaping on or off a tractor that is in motion.

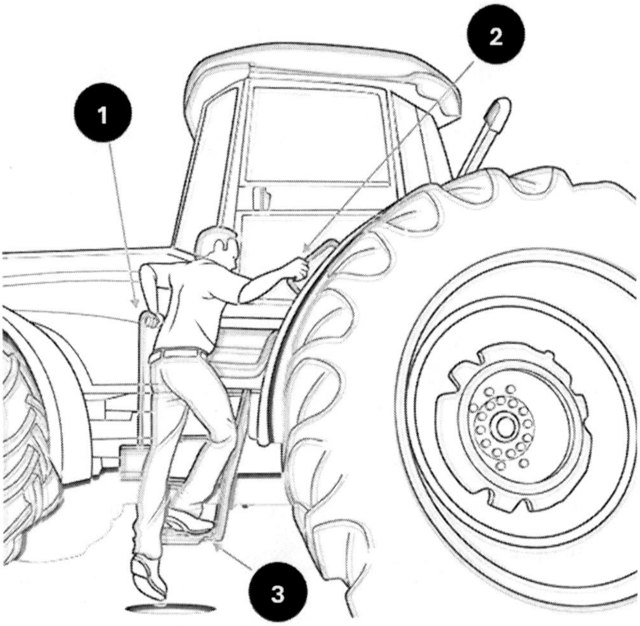

Figure 23: Maintaining 3 points of contact climbing in or out of the tractor.

Before starting the engine, meticulously check the vicinity of the tractor to ensure no individuals are nearby. Control the tractor exclusively from the seated position, maintaining three points of contact—two hands and one foot—when mounting or dismounting. Regularly clean tractor steps to prevent slips and falls, prioritizing safety during ingress and egress.

After completing the pre-start checks and starting the tractor, additional operational checks should be made to ensure everything is functioning properly before using the tractor. These checks typically include:

1. Engine Performance: Listen for any unusual noises or vibrations from the engine that may indicate mechanical issues. Ensure the engine starts smoothly and runs consistently at the correct RPM.

2. Hydraulic Systems: Test the hydraulic systems by operating any hydraulic controls or attachments. Check for proper function, smooth operation, and any signs of leaks in the hydraulic lines or fittings.

3. Steering and Brakes: Test the steering by turning the wheel in both directions to ensure it responds smoothly and without excessive play. Check the brakes by applying them gently to ensure they engage properly and hold the tractor in place.

4. Transmission and Gear Shift: Shift through all gears and test the transmission to ensure it engages smoothly and without hesitation. Check for any abnormal noises or grinding sounds during gear shifts.

5. Power Take-Off (PTO) System: Engage the PTO and test any attached implements to ensure they operate correctly and without unusual vibrations or noises. Check for proper engagement and disengagement of the PTO.

6. Lights and Signals: Once again, test all lights, including headlights, taillights, brake lights, turn signals, and hazard lights, to ensure they are still functional after starting the engine.

7. Temperature Gauges: Monitor temperature gauges for the engine coolant and hydraulic fluid to ensure they remain within safe operating ranges as the engine warms up.

8. Controls and Instruments: Double-check all controls, switches, and gauges to ensure they are still functioning correctly after starting the engine. Verify that all indicators and warning lights are operational.

By performing these additional operational checks after starting the tractor, operators can confirm that all systems are functioning properly and address any issues before using the tractor in actual operation. This helps ensure the safety and reliability of the equipment while minimizing the risk of accidents or breakdowns during use.

Unless within a soundproof cab, utilize hearing protection to safeguard against prolonged exposure to loud noises. When hitching loads, utilize the drawbar exclusively, ensuring it's set on a three-point hitch and not elevated beyond the fixed drawbar. Secure the drawbar pin appropriately, selecting the right size for the task. Employ counterweights, such as front-end or wheel weights, when necessary to enhance tractor stability.

Operate the clutch gradually at all times, exercising caution and precision. Before servicing equipment, unclogging, or making adjustments, disengage the power take-off (PTO), and switch off the engine. Adapt your operating speed according to surface conditions, prioritizing safety above all else.

During load transportation, maintain the loader bucket at a low position to enhance stability and prevent accidents. Take regular breaks during extended work periods, remaining vigilant of fatigue, and refrain from operating machinery under the influence of alcohol, drugs, or any impairing substances.

During operations:

- Mounting and Dismounting: Always approach and dismount from the left side of the tractor to avoid accidental contact with controls. This practice enhances safety during entry and exit.

- Protective Structures: Ensure that the tractor is equipped with an approved cab or a roll-over protective structure (ROPS) for added safety. When ROPS is present, use the seatbelt provided to secure yourself while operating the tractor.

- Safety Gear: Wear appropriate hearing protection while operating tractors, as not all cabs provide soundproofing. Additionally, consider the installation of a fall-on protective structure (FOPS) if there is a risk of falling objects.

- Key Management and Maintenance: Remove starter keys from tractors when not in use to prevent unauthorized operation. Maintain an up-to-date maintenance schedule and adhere to safe maintenance and jacking procedures to ensure equipment reliability.

- Operator Training and Comfort: Ensure that tractor operators are adequately trained for the specific tasks they will be performing. Adjust the seat position to ensure that all controls are within easy reach and that the operator can operate comfortably.

- Safety Precautions While Operating: Drive at controlled speeds to maintain control in unpredictable situations. Exercise caution when turning or applying brakes, especially on uneven terrain. Be vigilant for obstacles such as ditches, rocks, and depressions, and adjust driving techniques accordingly.

- Safe Towing Practices: Attach implements according to manufacturer instructions, using provided mounting points. Regularly inspect safety pins on towed implements and ensure all guards are in place before operation.

- Preventing Strain Injuries: Adjust the tractor seat for optimal back support and comfort. Take short breaks periodically to prevent strain injuries, and plan for future tractors to include features such as low steps, hand grips, and ample cab space for safe mounting and dismounting. Avoid jumping down from the tractor and utilize provided foot and handholds for safer dismounting.

Before utilizing any implements or attachments, it is essential to familiarize yourself with the manufacturer's safety manual thoroughly. This manual provides critical guidelines for safe operation and maintenance procedures. Prioritize safety by using only equipment that is in optimal condition. Ensure all guards and shields are intact and operational, especially around rotating components like power take-off shafts and gearboxes.

Maintain a safe distance from moving or rotating equipment, ideally operating from the tractor seat, and ensure bystanders are at least 6 meters (20 feet) away. Familiarize yourself with emergency procedures to quickly stop the tractor, engine, and attachment if necessary. Follow shutdown protocols, waiting for all moving parts to cease before dismounting the tractor or approaching the attachment.

When using post drivers, prioritize safety by ensuring all shields and guards are properly in place and functioning. Use a safety fork or guide to stabilize the post, reducing the risk of injury if it breaks or slips. Lower the hammer and switch off the engine before making adjustments or performing maintenance tasks. Securely lower and lock the hammer to prevent accidental movement, and never place any body part between the hammer and the post.

For auger operations, avoid entanglement hazards by keeping hair, clothing, hands, and feet away from the auger. Maintain a clear work area to prevent slips or falls and exercise caution when unclogging a stuck auger, shutting off the engine, and disengaging power before attempting to clear the blockage. Ensure all guards are correctly installed and functional, and be vigilant of overhead powerlines during movement or towing. Avoid using hands or feet to push material away from the auger to prevent injury.

When operating mowers and blades, prioritize safety by ensuring all guards are in place and the engine is shut off before making repairs. Understand how the blades operate and their location, keeping hands clear during operation and inspecting blades regularly, especially after hitting hard objects. Maintain a safe distance of approximately 92 meters (300 feet) from mowing operations for bystander safety and avoid mowing grass on slopes steeper than three to one incline, as advocated by Canadian Centre of Occupational Health and Safety (2024). Additionally, cut on the upside of a slope to maintain tractor stability and refrain from removing materials while the blade is running. Replace damaged or worn mower blades promptly to ensure safe operation.

When using attachments, it's crucial to prioritize safety measures. Begin by thoroughly reviewing and comprehending the safety manual provided by the manufacturer. This manual serves as a guide for safe operation and maintenance procedures. Always ensure that the equipment is in optimal condition before use, as any malfunction or damage could lead to hazardous situations. Check that all guards and shields are securely in place and functioning correctly, particularly around rotating or spinning components such as power take-off shafts and gearboxes.

Hydraulic systems play a vital role in many attachments, so it's essential to inspect all hydraulic lines for any signs of wear, damage, or leaks. Secure connections are crucial to prevent accidents and maintain operational efficiency. Additionally, consider adding front counterweights to tractors equipped with rear-mounted attachments to enhance steering control and overall stability during operation.

Proper installation of attachments is paramount to ensure safe usage. Utilize the correct brackets and follow the manufacturer's recommendations regarding bolt sizes and types for mounting equipment securely.

Employ appropriate tools such as jacks, hoists, or blocks to position equipment safely without risking injury.

Be mindful of potential crushing and entanglement points, and never position yourself between a tractor and an implement. Take precautions to avoid accidents by fastening safety locks and catches securely. When attaching the hitch, align the hitch holes from the tractor seat, and always switch off the engine before dismounting to avoid any potential crush points.

Under no circumstances should you crawl under anything supported solely by a hydraulic lift. Always use stable supports such as blocks or jack stands to prevent accidents. Regularly inspect and maintain all guards, including master and integral shields, ensuring they are free from damage. Any damaged guards or shields should be promptly repaired or replaced to maintain a safe working environment.

When operating the Power Take-Off (PTO), it's crucial to be keenly aware of its inherent dangers. The rapid spinning of the PTO can easily entrap individuals and pose serious risks, particularly to loose clothing or long hair.

When attaching or detaching PTO-driven equipment, strict safety measures must be followed:

It starts with familiarizing yourself with the manufacturer's safety manual, ensuring a comprehensive understanding of recommended procedures. Only deploy equipment that is in optimal condition, confirming that all guards or shields are securely in place and fully functional. Guards surrounding power take-off shafts, gearboxes, and other rotating components are of utmost importance.

Prior to attachment, lower hydraulics, shut off the tractor engine, and engage the parking brake. Safely hitch the tractor to the implement, ensuring universal joints are correctly aligned when connecting the shaft. Avoid wearing loose clothing, tie back long hair, and refrain from dangling shoelaces to minimize entanglement risks.

Maintain a safe distance from moving or rotating equipment, whenever feasible operating from the tractor seat, with bystanders positioned at least 6 meters (20 feet) away. Never remove shields from the PTO shaft, ensuring that spinner/integral shields rotate freely.

Select the appropriate size drive for the machine being powered and match the correct PTO speed accordingly. Avoid stepping over a rotating shaft, as even those with guards remain hazardous. Instead, walk around the equipment to mitigate risks.

Understand emergency procedures thoroughly, knowing how to swiftly stop the tractor, engine, and attachment if necessary. Follow shutdown protocols diligently, waiting for all moving parts to come to a complete halt before dismounting the tractor or approaching the attachment. When the PTO is not in use, disconnect it to prevent inadvertent activation and potential accidents.

Start by thoroughly reviewing, comprehending, and adhering to the owner's operating manual. Ensure you undergo proper training to effectively and safely operate a tractor. Conduct a comprehensive walk around check before initiating tractor operation or driving any vehicle.

When utilizing public roadways, ensure your tractor complies with highway regulations. Exercise caution and remain vigilant for potential hazards, including stumps, logs, rocks, holes, ditches, machinery, debris, buildings, and overhead powerlines.

Adjust your speed accordingly in various situations, such as when turning, navigating near obstacles like ditches, buildings, or trees, operating on slopes or rough terrain, pulling heavy or unstable loads, or driving on challenging surfaces like rough, soft, or slick ground. Take particular care when visibility is limited due to plants or darkness.

Here are some actions to avoid:

Firstly, refrain from using a tractor unless you have received proper training and understand how to operate it safely. Avoid allowing another person to ride on a tractor unless it is specifically designed to accommodate passengers, equipped with features like instructor seats, roll-over protection structures (ROPS), and safety belts.

Do not attempt to mount or dismount a moving tractor, as this poses a significant risk of injury. Additionally, avoid overloading the tractor, adhering strictly to the manufacturer's instructions regarding weight limits.

Exercise caution when starting the engine, ensuring you do not do so while standing beside the tractor. Furthermore, avoid driving at excessive speeds that cause the tractor wheels to bounce, which can compromise control and stability.

Avoid transporting loads that are poorly stacked, as they may shift or fall, posing hazards to both the operator and bystanders. Refrain from driving a tractor with a greasy or slippery steering wheel, as this can impede your ability to control the vehicle safely.

Never idle a tractor inside a building due to the potential accumulation of harmful fumes. Lastly, avoid driving a tractor directly towards anyone standing in front of a fixed object such as a building, wagon, or implement, as this can lead to collisions or accidents.

With regard to garden tractor operations, begin by thoroughly reviewing, comprehending, and adhering to the instructions outlined in the manufacturer's operating manual. It's essential to undergo proper training to safely operate a tractor. Familiarize yourself with operating the equipment and safely using attachments, ensuring you understand the function and location of all controls. Understand the tractor's speed capabilities, its ability to navigate slopes or rough terrain, as well as its braking and steering characteristics.

Before starting work, check the oil level and refuel the engine while it's cool. Ensure all shields, guards, and safety devices are in place and functional, such as those for power take-off, mower input drivelines, drive belts, chains, and gears. Maintain the tractor in optimal working condition by replacing or tightening any loose or damaged parts.

When operating the garden tractor, take the following precautions:

- Verify that the tractor is in neutral gear and disengage any attachment clutches before starting the engine, particularly if starting the tractor in an enclosed space to prevent carbon monoxide buildup. Conduct a pre-operational check of the tractor, including testing the brakes and ensuring the gas throttle functions correctly.

- Always check behind you before engaging reverse and drive the tractor up and down slopes for greater stability. Adjust speed accordingly, especially on slopes, sharp curves, or slippery surfaces. Use appropriate counterweights as recommended by the manufacturer for front or rear-mounted attachments.

- Before making any adjustments, turn off the machine, shift to neutral gear, set the brakes, and wait for all moving parts to stop. When leaving the tractor unattended, always turn off the machine, remove the key, and disengage the power take-off (PTO).

- Additionally, ensure the tractor receives regular servicing from a qualified mechanic, and take regular breaks to alleviate equipment vibration. Properly adjust the seat and steering wheel for comfort and accessibility.

Here are some actions to avoid when using a garden tractor:

- Refrain from riding on the tractor's hood or drawbar, and do not allow others to ride on the tractor. Avoid sharp turns, holes, ditches, or embankments that may cause the equipment to overturn. Park tractors in safe locations away from public endangerment.

- Do not tamper with or remove safety attachments, machine

guards, or safety labels, and use attachments specifically designed for the machine and task. Avoid clearing or unclogging the mower while the blades are in motion, and do not drive over gravel or rocks with rotating blades. Always ensure the tractor is attended to properly by turning off the power and removing the ignition key when leaving it unattended. Lastly, do not operate the tractor with the PTO running if it's not being used.

Using an auger on a tractor typically involves several steps to ensure safe and efficient operation. Here's a general guide on how to use an auger on a tractor:

- Preparation: Before attaching the auger to the tractor, ensure that both the tractor and the auger are in good working condition. Check for any signs of damage or wear on the equipment. Also, make sure that the auger is suitable for the type of soil or material you will be drilling into.

- Attach the Auger: Position the tractor on level ground and engage the parking brake. Align the tractor's three-point hitch with the attachment points on the auger. Use the tractor's hydraulic system to lift the auger into position and secure it using the attachment pins.

- Adjust the Auger: Depending on the depth and angle required for drilling, adjust the angle and height of the auger using the tractor's hydraulic controls. Ensure that the auger is securely attached and properly aligned with the ground.

- Start the Tractor: Start the tractor's engine and allow it to warm up to operating temperature. Ensure that all safety mechanisms are engaged, and the tractor is in gear.

- Engage the Auger: Slowly engage the power take-off (PTO) to start the auger's rotation. Begin drilling into the soil or material at a slow and steady pace, gradually increasing speed as needed. Pay attention to the tractor's engine RPM and avoid overloading the auger.

- Monitor Progress: As you drill, keep an eye on the depth and angle of the hole, as well as the performance of the tractor and auger. Adjust the speed and angle of the auger as necessary to maintain a consistent drilling process.

- Complete the Task: Continue drilling until you have reached the desired depth or completed the task at hand. Once finished, disengage the PTO and raise the auger using the tractor's hydraulic system.

- Detach the Auger: Lower the auger to the ground and disengage the attachment pins to detach it from the tractor. Carefully store the auger in a safe location, ensuring it is properly secured to prevent accidents or damage.

- Maintenance: After using the auger, perform any necessary maintenance tasks, such as cleaning and lubricating moving parts, inspecting for damage, and storing it in a dry and secure area.

Figure 24: Kubota tractor with Befco Auger on three-point hitch. User:SB_Johnny, CC BY-SA 3.0 , via Wikimedia Commons.

Using a front blade on a tractor is a practical method for various tasks, including snow removal, leveling soil, or clearing debris. As a general guide on how to use a front blade on a tractor:

1. Preparation: Before attaching the front blade, ensure that both the tractor and the blade are in good working condition. Check for any signs of damage or wear on the equipment. Also, make sure that the blade is suitable for the task at hand.

2. Attach the Blade: Position the tractor on level ground and engage the parking brake. Align the tractor's front loader attachment with the mounting points on the blade. Secure the blade to the loader arms using the attachment pins or any locking mechanisms provided.

3. Adjust the Blade: Depending on the task you're performing, adjust the angle and height of the blade using the tractor's hy-

draulic controls. Ensure that the blade is securely attached and properly aligned with the ground.

4. Start the Tractor: Start the tractor's engine and allow it to warm up to operating temperature. Ensure that all safety mechanisms are engaged, and the tractor is in gear.

5. Engage the Blade: Once you're ready to start, slowly engage the tractor's hydraulic system to lower the blade to the ground. Adjust the angle of the blade as needed to achieve the desired cutting or pushing angle.

6. Perform the Task: Depending on your specific job, you can use the front blade to push or scrape material, level uneven surfaces, or clear snow. Drive the tractor forward or backward as necessary, adjusting the blade angle and height to suit the terrain and task.

7. Monitor Progress: As you work, keep an eye on the performance of the tractor and blade. Adjust the blade angle, height, and position as needed to ensure efficient and effective operation.

8. Complete the Task: Once you've finished the job, disengage the hydraulic system and raise the blade back into its transport position. Secure any locking mechanisms or attachment pins to ensure the blade stays in place during transport or storage.

9. Detach the Blade: If you no longer need the front blade attached to the tractor, follow the manufacturer's instructions to safely detach it. Lower the blade to the ground, disengage any locking mechanisms, and remove the attachment pins.

10. Maintenance: After using the front blade, perform any necessary maintenance tasks, such as cleaning and lubricating moving

parts, inspecting for damage, and storing it in a dry and secure area.

Figure 25: Tractor set up for clearing snow. WrS.tm.pl, Public domain, via Wikimedia Commons.

Using a forklift attachment on a tractor can significantly enhance its versatility, allowing it to lift and transport heavy loads with ease. To use a forklift attachment on a tractor:

1. Preparation: Before attaching the forklift, ensure that both the tractor and the attachment are in good working condition. Inspect the forklift for any signs of damage or wear, and ensure that it is suitable for the task at hand.

2. Attach the Forklift: Position the tractor on level ground and engage the parking brake. Align the tractor's front loader attachment with the mounting points on the forklift. Secure the forklift to the loader arms using the attachment pins or any locking

mechanisms provided.

3. Adjust the Forks: Depending on the size and weight of the load you'll be lifting, adjust the position and width of the forks using the forklift's hydraulic controls. Ensure that the forks are securely attached and properly aligned for safe lifting.

4. Start the Tractor: Start the tractor's engine and allow it to warm up to operating temperature. Ensure that all safety mechanisms are engaged, and the tractor is in gear.

5. Engage the Forklift: Once you're ready to start lifting, slowly engage the tractor's hydraulic system to lower the forks to the ground. Position the forks under the load, ensuring that they are centred and stable.

6. Lift the Load: Gradually raise the forks using the tractor's hydraulic controls until the load is clear of the ground. Use caution and maintain a slow and steady pace to prevent the load from shifting or falling.

7. Transport the Load: Drive the tractor forward or backward as needed to transport the load to its destination. Keep the load balanced and secure, and avoid sudden movements or sharp turns that could cause instability.

8. Lower the Load: Once you've reached the desired location, carefully lower the forks to the ground using the tractor's hydraulic controls. Ensure that the load is placed securely and evenly on the ground before releasing it from the forks.

9. Detach the Forklift: If you no longer need the forklift attached to the tractor, follow the manufacturer's instructions to safely detach it. Lower the forks to the ground, disengage any locking

mechanisms, and remove the attachment pins.

10. Maintenance: After using the forklift attachment, perform any necessary maintenance tasks, such as cleaning and lubricating moving parts, inspecting for damage, and storing it in a dry and secure area.

Figure 26: Landini tractor with fork attachment. Bob Adams from Amanzimtoti, South Africa, CC BY-SA 2.0 , via Wikimedia Commons.

Using a rotary hoe attachment on a tractor can help in preparing soil for planting by breaking up and aerating the ground. This involves:

1. Preparation: Before attaching the rotary hoe, ensure that both the tractor and the attachment are in good working condition. Check for any signs of damage or wear on the equipment, and ensure that the rotary hoe is suitable for the type of soil and terrain you will be working on.

2. Attach the Rotary Hoe: Position the tractor on level ground and engage the parking brake. Align the tractor's three-point hitch with the attachment points on the rotary hoe. Use the tractor's hydraulic system to lift the hoe into position and secure it using

the attachment pins.

3. Adjust the Depth: Depending on the depth of cultivation required, adjust the height of the rotary hoe using the tractor's hydraulic controls. Lower the hoe to the desired depth, ensuring that it is set evenly across the entire width of the attachment.

4. Start the Tractor: Start the tractor's engine and allow it to warm up to operating temperature. Ensure that all safety mechanisms are engaged, and the tractor is in gear.

5. Engage the Rotary Hoe: Slowly engage the power take-off (PTO) to start the rotation of the hoe blades. Begin driving the tractor forward at a slow and steady pace, allowing the rotating blades to penetrate the soil.

6. Monitor Progress: As you cultivate the soil, keep an eye on the depth and quality of the cultivation. Adjust the speed and depth of the rotary hoe as necessary to achieve the desired results.

7. Work in Passes: For larger areas, it may be necessary to make multiple passes with the rotary hoe to ensure thorough cultivation. Overlapping each pass slightly can help to achieve even coverage and prevent missed spots.

8. Turn with Caution: When reaching the end of a row or turning the tractor, use caution to avoid damaging the rotary hoe or the tractor. Lift the hoe out of the ground before making sharp turns, and lower it back down once you are ready to continue cultivating.

9. Complete the Task: Continue cultivating the soil until you have covered the entire area or achieved the desired level of cultivation. Once finished, disengage the PTO and raise the rotary hoe

using the tractor's hydraulic system.

10. Detach the Rotary Hoe: Lower the rotary hoe to the ground and disengage the attachment pins to detach it from the tractor. Carefully store the rotary hoe in a safe location, ensuring it is properly secured to prevent accidents or damage.

11. Maintenance: After using the rotary hoe attachment, perform any necessary maintenance tasks, such as cleaning and lubricating moving parts, inspecting for damage, and storing it in a dry and secure area.

Using spraying equipment on a tractor allows for the efficient application of fertilizers, pesticides, herbicides, and other chemicals to crops.

Figure 27: Tractor crop-spraying. Andrew Tatlow, CC BY-SA 2.0 , via Wikimedia Commons.

To use spraying equipment on a tractor:

1. Preparation: Before attaching the spraying equipment, ensure that both the tractor and the sprayer are in good working condition. Check for any signs of damage or wear on the equipment, and ensure that the sprayer is filled with the appropriate chemicals according to the application requirements.

2. Attach the Sprayer: Position the tractor on level ground and engage the parking brake. Align the tractor's three-point hitch with the attachment points on the sprayer. Use the tractor's hydraulic system to lift the sprayer into position and secure it using the attachment pins.

3. Adjust Spray Settings: Depending on the type of chemical being applied and the desired coverage, adjust the spray settings on the sprayer. This may include adjusting the spray nozzles for the desired spray pattern, adjusting the pressure settings for optimal coverage, and calibrating the flow rate to ensure accurate application.

4. Start the Tractor: Start the tractor's engine and allow it to warm up to operating temperature. Ensure that all safety mechanisms are engaged, and the tractor is in gear.

5. Engage the Sprayer: Once the tractor is ready, engage the power take-off (PTO) to activate the sprayer's pump. Begin driving the tractor forward at a slow and steady pace, allowing the sprayer to emit a continuous spray of chemicals onto the crops.

6. Monitor Application: As you spray, monitor the application rate and coverage to ensure uniform distribution of the chemicals. Adjust the tractor's speed and the sprayer's settings as necessary to achieve the desired application rate and coverage.

7. Maintain Distance: Maintain a consistent distance between the

sprayer and the crops to ensure even coverage and minimize drift. Avoid spraying in windy conditions or during periods of high humidity to prevent off-target drift of chemicals.

8. Turn with Caution: When reaching the end of a row or turning the tractor, use caution to avoid damaging the sprayer or the crops. Lift the boom or turn off the sprayer before making sharp turns, and resume spraying once you are back on course.
9. Complete the Application: Continue spraying until you have covered the entire area or achieved the desired application rate. Once finished, disengage the PTO and lower the sprayer using the tractor's hydraulic system.
10. Rinse and Clean: After spraying, thoroughly rinse the sprayer's tanks, lines, and nozzles with clean water to prevent chemical buildup and contamination. Dispose of any leftover chemicals properly according to local regulations.
11. Maintenance: After using the spraying equipment, perform any necessary maintenance tasks, such as cleaning and lubricating moving parts, inspecting for damage, and storing it in a dry and secure area.

Figure 28: Spraying pesticides. Maasaak, CC BY-SA 4.0, via Wikimedia Commons.

Finishing up Tractor Operations

Upon completing operations, engage the parking brake, shift into park or neutral, and remove the ignition key before leaving the tractor. These precautions contribute to overall safety and prevent unintended movement or accidents. Specifically, to park up, shut down, secure, and carry out a post-operational inspection of a tractor and its attachments in line with workplace procedures, follow these steps:

1. **Park Up**:

    - Bring the tractor to a flat, stable surface away from traffic and obstacles.
    - Ensure the tractor is in a safe parking location, away from any hazards.
    - Engage the parking brake to secure the tractor in place.

- Place the transmission in neutral and disengage the power take-off (PTO).

2. **Shut Down**:
   - Reduce the engine speed to idle for a few minutes to allow it to cool down.
   - Turn off any attached equipment or implements.
   - Lower any raised attachments, such as loaders or mowers, to the ground.
   - Turn off the engine using the ignition switch or key.
3. **Secure**:
   - Remove the key from the ignition to prevent unauthorized use.
   - Apply any additional security measures required by workplace procedures, such as locking the cab or removing detachable parts.
4. **Post-Operational Inspection**:
   - Visually inspect the tractor and attachments for any signs of damage, leaks, or wear and tear.
   - Check the tyres for proper inflation, tread wear, and damage.
   - Inspect all fluid levels, including engine oil, hydraulic fluid, coolant, and fuel, and top up as necessary.
   - Verify that all lights, signals, and indicators are functioning correctly.

- Test the brakes, steering, and other controls to ensure they operate smoothly and effectively.
- Check for any loose or missing bolts, nuts, or fasteners on attachments.
- Clean and lubricate any moving parts or components as needed.
- Review any operator logs or maintenance records to ensure compliance with workplace procedures and regulatory requirements.

Chapter Three

Cultivators

A cultivator is a type of agricultural implement used for soil cultivation and weed control in farming. It typically consists of a series of curved or straight blades, called tines, attached to a frame. The tines can be made of metal or durable plastic and are designed to break up soil, remove weeds, and aerate the ground in preparation for planting.

Cultivators come in various sizes and configurations, ranging from handheld tools for small-scale gardening to large tractor-mounted implements for commercial farming operations. Handheld cultivators are often used in home gardens or small-scale farming to manually cultivate soil between rows of plants or in tight spaces where larger equipment cannot reach.

Cultivators serve as indispensable implements in diverse agricultural contexts, assisting in tasks like tilling, weeding, and seedbed preparation. They can be operated manually, by animals, or through machinery (Rajput, 2023). In agriculture, cultivators play a crucial role in soil preparation, weed management, and fostering ideal conditions for plant growth.

On the other hand, tractor-mounted cultivators are more commonly used in larger agricultural settings. These cultivators are attached to the back of a tractor and drawn through the field, effectively cultivating large swaths of land in a single pass. They may feature multiple rows of tines arranged in a row or staggered pattern, allowing for efficient soil cultivation and weed control.

Figure 29: Perfecta II field cultivator fitted to a tractor. Dwight Sipler, CC BY 2.0 , via Wikimedia Commons.

Cultivators play a vital role in preparing the soil for planting by breaking up compacted soil, incorporating organic matter, and creating a loose, friable seedbed. Additionally, they help control weeds by uprooting or burying them, reducing competition for nutrients and water and promoting healthier crop growth.

Cultivation refers to the practice of preparing and working the soil for planting crops. It involves various techniques aimed at improving soil structure, fertility, and drainage to create an optimal environment for plant growth. Cultivation methods may include tilling, plowing, harrowing, and hoeing, among others, depending on the specific needs of the crops being grown and the conditions of the soil.

The primary objectives of cultivation are to:

1. Break up compacted soil: Soil compaction can occur due to factors such as heavy machinery, foot traffic, or natural settling. Cultivation helps loosen compacted soil, allowing roots to penetrate more easily and access nutrients and water.

2. Control weeds: Weeds compete with crops for nutrients, water, and sunlight, reducing overall yield and quality. Cultivation helps suppress weed growth by uprooting or burying weed seeds and seedlings, preventing them from germinating and establishing themselves.

3. Incorporate organic matter: Organic matter, such as compost or manure, improves soil structure, fertility, and microbial activity. Cultivation helps incorporate organic matter into the soil, enriching it with essential nutrients and promoting healthy microbial communities.

4. Improve drainage: Excess water can lead to waterlogging, root rot, and other problems detrimental to plant health. Cultivation helps improve soil drainage by breaking up compacted layers and creating channels for water to flow through, reducing the risk of waterlogging and promoting healthier root development.

5. Create a seedbed: Cultivation creates a suitable seedbed for planting by preparing the soil surface to a fine, crumbly texture. A well-prepared seedbed provides optimal conditions for seed germination, root growth, and seedling establishment.

A crop cultivator is a specialized agricultural tool or machine utilized for soil cultivation, aiming to ready the soil for planting and nurturing crops. This equipment exists in diverse variants, ranging from tractor-mounted cultivators to handheld devices, and even specialized models tailored for particular crops or farming techniques.

A tractor cultivator represents a specific type of agricultural apparatus affixed to a tractor, primarily employed for soil cultivation within farming endeavours. Engineered to be towed by a tractor, it serves various purposes such as seedbed preparation, weed management, and integration of fertilizers into the soil.

Cultivators are typically categorized into two main groups: Primary cultivators and Secondary cultivators (Searle, 2018).

Figure 30: Cultivating with a Challenger MT765C tractor and a Simba SL500 cultivator/disc harrow/roller combination. David Wright, CC BY 2.0 , via Wikimedia Commons.

Primary Cultivators: Primary cultivators are primarily utilized on uncultivated soil or heavy ground requiring thorough preparation before seeding. Examples of primary cultivation machines include:

1. Mouldboard Ploughs (See Figure 31): Mouldboard ploughs not only loosen the soil but also invert it, effectively eliminating germinated weeds in the process.

2. Disc Ploughs/Cultivators (see Figure 30): Ideal for primary cultivation and land clearing tasks, disc ploughs feature leading

disc blades that initially rip the ground, followed by rear discs that further chop it down. In some cases, disc cultivators can serve as both primary and secondary cultivators due to their dual functionality.

Figure 31: Example of a Mouldvoard Plough. Hjart, CC BY-SA 4.0>, via Wikimedia Commons.

Secondary Cultivators: Secondary cultivators are employed after the initial soil preparation by primary cultivators or as the sole method of cultivation for lighter ground. These cultivators are generally lighter in weight and require less tractor power. Examples of secondary cultivators include:

1. Tine Cultivators (See Figure 29): Tine cultivators resemble scaled-down versions of chisel ploughs, featuring rows of tines that loosen the upper soil layers without deeply disrupting the ground structure. They are ideal for final seed bed preparation, leaving a smooth finish and often used alongside crumbler rollers or finger tines for even soil leveling. Tine cultivators can cover large areas with lower power requirements compared to other secondary cultivators, contributing to their popularity.

They are available in various sizes and mounting options, from 3-point linkage to fully trailing configurations.

2. Combination Cultivators: Combination cultivators are versatile machines capable of both primary and secondary cultivation tasks. These machines may incorporate discs, heavier tine cultivators, or a combination of both, providing flexibility in soil preparation methods.

The primary and secondary cultivators differ in their functions, applications, and the stage of soil preparation they are used for in agricultural operations. Here are the main differences between primary and secondary cultivators:

1. Purpose:

    - Primary cultivators: These are used for initial soil preparation on uncultivated land or heavy ground that requires thorough breaking up and inversion of soil layers. Their primary purpose is to prepare the soil for seeding by loosening it and eliminating weeds.

    - Secondary cultivators: Secondary cultivators are employed after primary cultivation to refine the soil further, create a finer seedbed, and prepare it for planting. They are used to level the soil surface, remove clods, and create an optimal environment for seed germination and plant growth.

2. Soil Penetration:

    - Primary cultivators: These cultivators penetrate deeper into the soil, breaking up compacted layers and preparing the ground for seeding. They may invert the soil, bringing nutrient-rich subsoil to the surface.

- Secondary cultivators: Secondary cultivators work closer to the soil surface, targeting shallow soil layers to refine the seedbed. They focus on leveling the soil and breaking up smaller clods without deeply disturbing the soil structure.

3. Equipment Design:

- Primary cultivators: Equipment like mouldboard ploughs and disc ploughs/cultivators are commonly used as primary cultivators. They are designed with heavy-duty components to withstand the forces involved in breaking up tough soil.

- Secondary cultivators: Tine cultivators and combination cultivators are examples of secondary cultivators. They are lighter in weight and often feature adjustable settings for finer soil preparation and seedbed creation.

4. Power Requirements:

- Primary cultivators: Due to their deeper soil penetration and heavier construction, primary cultivators typically require more power from the tractor or machinery to operate effectively.

- Secondary cultivators: Secondary cultivators are lighter and require less power compared to primary cultivators. They can be operated with smaller tractors or machinery, making them more versatile for different farming operations.

5. Stage of Soil Preparation:

- Primary cultivators: These are used at the beginning of the soil preparation process to break up uncultivated or heavily compacted soil and prepare it for planting.

- Secondary cultivators: Secondary cultivators are used after primary cultivation to refine the soil surface, create a smooth seedbed, and ensure optimal conditions for seed germination and crop establishment.

By understanding these differences, farmers can select the appropriate cultivator for each stage of soil preparation in their agricultural operations, ensuring efficient and effective crop production.

Cultivator and tiller are frequently interchanged terminologies, yet distinctions exist between them. Cultivators typically feature smaller blades intended to loosen the soil, whereas farm or garden tillers are equipped with larger blades designed to penetrate deeper into the soil, breaking up clumps and roots. Consequently, cultivators are more effective for smaller areas, whereas tillers are better suited for larger lands and fields (Yazid, 2023).

Planning and Preparing for Cultivator Operations

There are various types of cultivators, including gas-powered, electric, and handheld models. Selecting the right cultivator for the task includes (Yazid, 2023):

1. Surface Area: The extent of the area requiring cultivation determines the type of cultivator to choose. For vast expanses like open fields, tractor-mounted cultivators such as chisel plows or chain harrows are advisable, saving both time and effort. Conversely, for smaller areas, a manual handheld cultivator may suffice for the task.

2. Depth of Tillage: Cultivation involves breaking up the soil with a plow or tiller to facilitate planting. Handheld cultivators typically work well on shallow soils, whereas trailed cultivators like

disc harrows offer a wider range of depth options, depending on their ability to penetrate the soil deeply.

3. Soil Type: The nature of the soil influences the choice of cultivator. In rocky or boulder-rich terrain, a lighter cultivator is preferable for easier manoeuvrability and reduced risk of getting stuck. Conversely, for soft soils that are easy to break down, a heavy-duty cultivator capable of deeper penetration is ideal.

4. Power Source: Cultivators are powered by various sources, including gas and electricity. Electric cultivators may suffice for small farming areas, while larger areas may require gas-powered models for their increased power. Gas-powered cultivators range from lightweight models with a one-piece handlebar to heavy-duty disc harrows.

Manual cultivators are favoured by gardeners with smaller plots or fewer plants, finding them suitable for loosening the top two inches of soil and effectively managing surface weed growth, particularly in loose or compacted soil. Similarly, multi-farm hand cultivators cater to small areas or pastoral plots, equipped with rubber tyres for easy manoeuvrability and reduced soil compaction, alongside various adjustments and a depth control scale for customized use. Meanwhile, mini manual cultivators, widely used in orchards, fields, and mountainous terrain, offer versatility for tasks like weeding, loosening soil, or preparing farmland, featuring two wheels convertible into one for ease of use in narrow areas like roads, and a quick-release mechanism for part replacement as per farming needs.

Engine-powered cultivators, whether gas or electric, serve larger gardening needs by effectively preparing garden beds, breaking hard soil, or weeding between crop rows and aerating soil around plants. While gas-powered models offer more power albeit with higher maintenance needs, electric models are lightweight and user-friendly but may not handle tough soil or weeds as effectively. Electric cultivators, like the mini rotary cordless cultivator suitable for gardening and farming tasks such as soil cultivation or aerating, provide hassle-free operation without spark or gas exhaust, boasting a working depth of 180 mm and an idle speed of 150 rpm. Similarly, garden corded electric cultivators, powered by an AC motor, feature an "H type" handle and adjustable electric cord for extended working time compared to cordless models, equipped with 6×4 blades for digging generous holes in the ground.

Gas cultivators, including mini rotary cultivators ideal for various terrains and field tasks and farm gasoline cultivators renowned for portability and durability, cater to diverse farming needs. Mini rotary cultivators offer adjustable suspension for precise soil penetration and compatibility with multiple attachments, while farm gasoline cultivators boast wear-resistant wheels and optional accessories such as tillage plows or flip plows.

Figure 32: Mini rotary cultivator. Fallaner, CC BY-SA 4.0 , via Wikimedia Commons.

Additionally, cultivators like chain harrows, drag chain harrows, grass frame harrows, disc harrows, heavy-duty disc harrows, trailed offset disc harrows, and chisel plows cater to specific soil conditions and farming requirements, offering effective soil preparation, weed removal, and soil aeration. Each type features unique characteristics, from the simplicity and efficiency of chain harrows to the versatility and power of disc harrows and chisel plows, ensuring farmers can select the most suitable cultivator for their specific needs.

Planning and preparing for using a cultivator involves several key steps to ensure efficient and effective operation. Firstly, it's essential to assess cultivation needs by determining the specific tasks required, such as soil preparation, weed control, or seedbed preparation. Additionally, assessing the size of the area to be cultivated and the type of soil being worked with is crucial for proper planning.

Once the needs are assessed, selecting the right cultivator is the next step. Choosing the appropriate type based on the size of the area and specific requirements is important. Factors to consider include power source (manual, electric, gas-powered), size, and attachment options.

After selecting the cultivator, checking its equipment condition is vital. Inspecting for any damaged or worn-out parts that need replacement and ensuring all bolts, nuts, and connections are tight and secure is necessary for safe operation.

Preparing the area comes next, involving clearing the cultivation area of debris, rocks, or large obstacles that could obstruct the cultivator's operation. Removing weeds and vegetation from the surface helps prevent them from getting tangled in the cultivator blades.

Adjusting the cultivator's settings, such as depth and width, according to specific cultivation tasks and soil type is essential. Planning cultivation paths to minimize overlap and ensure thorough coverage of the entire area is also crucial, considering factors such as land layout, obstacles, and access points.

Checking weather conditions before starting cultivation is important to ensure optimal working conditions. Avoiding extremely wet or dry conditions helps maintain the performance and quality of cultivation.

Ensuring safety precautions, such as wearing appropriate personal protective equipment and following all safety guidelines outlined in the cultivator's manual, is necessary for safe operation.

Fuelling and performing any necessary maintenance tasks, such as oil checks or blade sharpening, on gas-powered cultivators is important to ensure smooth operation during cultivation.

Lastly, practicing operation, especially if unfamiliar with the cultivator, helps familiarize oneself with its controls and manoeuvring. Practicing in an open area before starting work on the cultivation site is recommended.

Cultivation, while essential for agricultural activities, poses various hazards that workers need to be aware of to ensure their safety. Some common hazards associated with cultivation include:

- Machinery Accidents: Working with cultivators, tractors, and other agricultural machinery can pose significant risks, including entanglement, crushing, and impact injuries. Accidents can occur due to improper use, lack of maintenance, or mechanical failures.

- Falls: Working on uneven terrain or slippery surfaces can increase the risk of falls, especially when operating machinery or walking on elevated platforms. Falls from height, such as off tractors or cultivators, can result in serious injuries.

- Chemical Exposure: The use of pesticides, herbicides, and fertilizers in cultivation exposes workers to hazardous chemicals. Inhalation, skin contact, or ingestion of these chemicals can lead to acute or chronic health effects, including respiratory issues, skin irritation, and poisoning.

- Musculoskeletal Injuries: Manual handling of heavy equipment, lifting, bending, and repetitive tasks associated with cultivation can lead to musculoskeletal injuries such as strains, sprains, and back problems. Improper lifting techniques and overexertion contribute to these injuries.

- Noise Exposure: Operating noisy machinery such as tractors, cultivators, and harvesters without proper hearing protection can lead to hearing loss and other auditory problems over time.
- Heat Stress: Working outdoors in hot and humid conditions during cultivation seasons can lead to heat-related illnesses such as heat exhaustion and heatstroke. Prolonged exposure to high temperatures without adequate hydration and rest can be dangerous.
- Biological Hazards: Working in fields exposes workers to various biological hazards, including bites from insects, snakes, or other pests, as well as exposure to allergens from plants and mold spores.
- Electrical Hazards: Contact with overhead power lines or faulty electrical equipment poses the risk of electric shock and electrocution, especially when operating tall machinery or irrigation systems near power sources.
- Confined Spaces: Certain cultivation activities may involve working in confined spaces such as silos or storage bins, which can pose risks of suffocation, entrapment, or engulfment if proper safety precautions are not followed.
- Weather-Related Hazards: Adverse weather conditions such as lightning, storms, high winds, or extreme temperatures can pose significant risks to workers during cultivation activities.

To mitigate these hazards, employers should provide adequate training, personal protective equipment (PPE), and safety protocols for workers. Regular equipment maintenance, proper supervision, and adherence to safety guidelines are essential to prevent accidents and injuries during cultivation operations.

Attaching a cultivator to a tractor involves several steps to ensure proper installation and safe operation. This includes:

1. Select the Right Cultivator: Choose a cultivator that is compatible with your tractor's size, hitch type, and power capabilities. Ensure that the cultivator is appropriate for the specific cultivation tasks you intend to perform.

2. Prepare the Tractor: Park the tractor on a flat, stable surface and engage the parking brake. Turn off the engine and remove the ignition key for safety. Inspect the tractor's hitch system to ensure it's clean, lubricated, and in good working condition.

3. Position the Cultivator: Position the cultivator behind the tractor in a suitable location for hitching. Align the cultivator's hitch tongue or frame with the tractor's hitch system. Ensure that the cultivator is centred and straight behind the tractor for even operation.

4. Lower the Implement: If the cultivator has hydraulic lift capabilities, lower the implement to the ground using the tractor's hydraulic controls. This will make it easier to align the cultivator with the tractor's hitch system.

5. Attach the Hitch: Manoeuvre the tractor's hitch system (such as a three-point hitch or drawbar) into position to align with the cultivator's hitch tongue or attachment points. Lower the tractor's hitch mechanism or drawbar until it engages securely with the cultivator's hitch.

6. Secure Attachment Pins: Insert and secure attachment pins or

locking mechanisms to fasten the cultivator to the tractor's hitch system. Ensure that all pins are properly inserted and locked in place to prevent accidental detachment during operation.

7. Connect Hydraulic Lines (if applicable): If the cultivator requires hydraulic power for raising, lowering, or angling, connect the hydraulic hoses from the cultivator to the tractor's hydraulic ports. Follow the manufacturer's instructions to ensure proper connection and avoid hydraulic leaks.

8. Check Attachment: Once the cultivator is hitched to the tractor, visually inspect the attachment points, hitch connections, and hydraulic lines (if applicable) to ensure everything is secure and properly aligned. Test the cultivator's hydraulic functions to verify smooth operation.

9. Adjust Hitch Height (if necessary): Use the tractor's hydraulic controls to adjust the height of the cultivator as needed for the desired cultivation depth. Refer to the cultivator's manual for recommended depth settings based on soil type and cultivation requirements.

10. Perform Safety Checks: Before operating the tractor-cultivator combination, double-check all connections, hitch pins, and safety features. Ensure that the cultivator is clear of any obstructions and that all safety shields and guards are in place.

11. Test Operation: Start the tractor's engine and test the cultivator's operation by raising and lowering it a few times to ensure smooth movement. Practice manoeuvring the tractor-cultivator combination in an open area to familiarize yourself with its handling characteristics before beginning work in the field.

Figure 33: Row-crop cultivator attached to tractor. Einboeck.official, CC BY-SA 4.0 , via Wikimedia Commons.

Cultivator Operations

Soil preparation is the initial step in the farming process, involving the manipulation of the soil to create an optimal environment for seed germination and plant growth. This process typically includes tilling or plowing the land to loosen the soil and establish a suitable seedbed.

Tilling or plowing serves several purposes. Firstly, it breaks up compacted soil, allowing air, water, and nutrients to penetrate deeper into the soil profile. This aeration of the soil promotes root development and improves overall soil structure. Secondly, it incorporates organic matter and soil amendments evenly throughout the soil, enhancing its fertility and nutrient content. Thirdly, tilling or plowing helps to control weeds by uprooting existing vegetation and burying weed seeds deeper in the soil, reducing their germination and growth.

The depth and method of tillage depend on various factors, including soil type, crop type, and farming practices. Soil type refers to the texture, composition, and structure of the soil, which can vary greatly from one location to another. Different soil types may require different depths and intensities of tillage to achieve the desired seedbed. For example, sandy soils may require shallower tillage compared to clay soils to prevent excessive disruption and loss of soil structure.

Crop type also influences soil preparation methods. Some crops may require a finely tilled seedbed with a smooth surface for optimal seed germination and root establishment, while others may tolerate rougher soil conditions. Additionally, certain crops have specific soil requirements, such as moisture retention or drainage, which may necessitate adjustments in tillage depth and method.

Farming practices, such as conventional or no-till farming, also play a role in soil preparation. Conventional tillage involves thorough soil disturbance through plowing or intensive tilling to prepare the seedbed. In contrast, no-till farming minimizes soil disturbance by planting seeds directly into untilled soil or into crop residues from previous harvests. The choice between conventional and no-till methods depends on factors such as soil erosion risk, water conservation goals, and long-term soil health considerations.

Before delving into conventional soil preparation methods, it's essential to understand the concept of no-till farming (also known as zero tillage or direct drilling). No-till farming is a cultivation technique aimed at growing crops without disturbing the soil through tilling (Farmbrite, 2022). This approach offers numerous advantages, such as reducing soil erosion, lowering irrigation requirements, minimizing the need for fertilizers and soil amendments, decreasing weed pressure, and preserving soil biology, nutrients, and overall health.

In no-till farming, planting typically occurs either by over-seeding or directly seeding into or over a previous cover crop. The cover crop serves as a natural mulch and compost for the new planting while providing additional weed suppression. Sometimes, farmers may opt to apply herbicides to the previous cover crop or utilize silage tarps to prepare the planting area. Silage tarps, large UV-coated plastic sheets, can be spread over a cover crop, suffocating and killing any weed seeds lying dormant within as little as three weeks. This process effectively protects the soil's microbiology and nutrients. Once complete, the bed is left ready for planting with minimal labour and enriched nutrient content (Farmbrite, 2022).

Traditionally, tillage has been the primary method for soil preparation on farms. However, an increasing number of farmers are recognizing the benefits of adopting no-till or low-till techniques, which involve minimal or shallow tillage, often using equipment like disc harrows. Each soil preparation method, whether tillage-based or no-till, comes with its own set of advantages and disadvantages, so it's crucial to research and understand which approach may work best for your specific operation.

In many regions, spring brings abundant moisture in the form of rain or snow, which can be beneficial for perennial plants and trees but poses challenges for soil preparation. Attempting to work overly wet soils is not only difficult but also leads to excessive soil compaction, hindering root growth and nutrient uptake. Before initiating any digging, plowing, or tilling activities, it's essential to ensure that the soil moisture levels have sufficiently decreased to manageable levels. Conducting a basic soil moisture test can help determine if the soil is ready for preparation. Scoop up a handful of soil and gently form it into a ball; then, drop the ball onto a hard surface. If the ball bounces or forms a puddle, the soil likely still contains too much moisture and requires more time to dry. Conversely, if the ball easily breaks apart, the soil is likely dry enough to begin preparation activities.

For farmers who utilized fall or winter cover crops to enhance soil fertility, adjust pH levels, or promote soil stability, mowing and incorporating these cover crops into the field is a crucial step before bed preparation and planting. Initially, the cover crop needs to be cut down using appropriate equipment such as tractors, walk-behind tractors, heavy-duty mowers, or scythes. Once the cover crop has been cut, it's beneficial to allow the plant matter to settle in place for a few days before incorporating it into the soil through tilling or other methods. This settling period allows for partial decomposition of the plant material. Following this, the cover crop and root material should be fully tilled or plowed into the field. Thoroughly breaking down the cover crop aids in the efficient absorption of nutrients into the soil and minimizes the risk of cover crop re-growth. It's advisable to allow the field to rest for 3-6 weeks after incorporating the cover crop before proceeding with bed preparation or seed sowing to facilitate maximum nutrient absorption into the soil.

Although soil sampling is commonly done in the fall to assess soil health post-harvest, obtaining an additional soil sample in the spring, especially after incorporating cover crops, can offer valuable insights into overall soil health and potential amendments needed.

Given that soil is a dynamic and diverse biological system, regular soil check-ups are essential. Think of soil samples as routine health check-ups for your soil, providing valuable information for better management practices.

Depending on your soil's health, nutrient levels, structure, or composition, you may need to introduce various amendments to create an optimal environment for your intended crops. Organic soil amendments not only provide nutrients but also offer additional benefits such as improving soil structure, enhancing water retention, increasing soil fertility, and reducing the need for labour-intensive tillage. However, it's crucial not to exceed adding more than 7.6-10.2 centimetres (3-4 inches) of compost or organic matter to the soil.

There are numerous types of soil amendments available, but we'll focus on some of the more common organic options (Farmbrite, 2022):

Organic Plant Matter: Incorporating cover crops into your soil is an effective way to enrich it with additional nutrients. Additionally, organic plant materials like leaves, straw, or grass clippings can be added. It's best to incorporate these amendments in the fall to allow sufficient time for decomposition over the winter.

Compost: Whether homemade or purchased from a local source, compost is an excellent source of organic matter and nutrients for the soil. Compost can also serve as a carrier for integrating other amendments such as gypsum, biochar, kelp, manures, azomite, humates, bone meal, sea minerals, boron, copper, zinc, etc. It's advisable to mix any additional amendments with compost a few weeks before incorporation.

Manure: Manure provides valuable nutrients and can be sourced from various livestock such as cattle, goats, or sheep. However, it's essential to let the manure age before use to prevent damage to plants and minimize the risk of introducing harmful bacteria. Typically, about 13.6-18.1 kilograms (30-40 lbs) of aged and composted manure are recommended per 9.3 square meters (100 sqft) of soil.

Green Manure: Growing nutrient-rich plants, cutting them, and allowing them to break down into the soil before tilling can improve soil health. Planting cover crops like rye or oats in the fall after soil preparation is an effective way to incorporate additional organic matter.

Gypsum or Sand: For heavy clay soils, adding gypsum or sand can help loosen the soil and make it more manageable. Adding 1.4-1.8 kilograms (3-4 lbs) of gypsum per 9.3 square meters (100 sqft) after fall harvest can enhance the workability of heavy clay soils in preparation for spring planting.

When opting for tillage, it's advisable to aim for a depth of at least 20-25 centimetres (8-10 inches). This practice helps to loosen the soil, break down existing plant matter, and facilitates deeper root penetration for your crops. Whether using a walk-behind tractor, rototiller, or traditional tractor, it's essential to till when the soil is dry to avoid difficulties and prevent additional compaction and soil damage. Preparing the soil in winter by spading or forking can also ease spring tilling.

Preparing beds and working the soil may be physically demanding, but numerous mechanized solutions are available to help prepare your soil for planting, regardless of its current condition. Depending on field conditions and crop types, different implements may be required for soil preparation. There are various types and configurations of plows designed to cater to different depths, aeration levels, and soil impacts. If considering purchasing a plow or tractor attachment, consulting with farm equipment retailers and seeking advice from friends and neighbours familiar with your specific needs, soil conditions, and crop plans is recommended. The following an overview of some common implements used for soil preparation (Farmbrite, 2022):

- Rototiller: Rototillers, available in various sizes and configurations, utilize rotating tines or discs to dig into the soil, loosen it, and break it up. They are particularly effective for breaking up compacted soils.

- Moldboard Plow: Moldboard plows feature a large curved blade pulled across the ground by a tractor. This blade digs deep into the soil, cutting a row and turning it over. The process not only loosens and aerates the soil but also leaves behind a planting trough.

- Reversible Garden Plow: Similar to moldboard plows but equipped with two or more reversible blades, reversible plows allow for creating various space furrows or mounds by mounting blades in different positions and directions. They are commonly used for deep tillage to turn over the upper layer of soil and bring up fresh nutrients.

- Chisel Plow: Chisel plows utilize a curved chisel or rod to aerate and loosen the soil without turning it over, leaving crop residue on the soil surface. Available in various configurations and widths, they are less damaging to soil health due to reduced impact.

- Disc Plow: Disc plows, pulled behind a tractor, feature multiple rows of rotating steel discs that effectively pulverize and break up compacted soil, making them suitable for rough ground.
- Subsoil Plow: Designed for breaking up hardpan layers beneath the soil surface, subsoil plows pull lower soil layers to the surface without turning them. This action improves soil aeration, minimizes compaction, and facilitates water penetration to disrupt hardpan layers.

Now that your field has been dried, amended, and tilled, it's time to ready it for planting. Fields are often prepared into beds to facilitate planting, maintenance, and harvesting, enhancing efficiency. Additionally, bed preparation allows for proper drainage away from plant roots, creates irrigation troughs, and promotes airflow.

Before forming beds, it's essential to ensure your growing area is as flat as possible. A level surface aids in easier germination for direct sowing and provides a smoother platform for transplantation. To achieve this, utilize a pull-behind rake attachment or a sturdy soil rake to flatten the field.

Once the ground is level, the next step is to create rows (furrows) spaced approximately every 36 inches, assuming your garden is of sufficient size. Beds in the US typically measure 36 inches in width and can extend up to 100 feet in length. While straight beds are aesthetically pleasing, they are not mandatory. Consider the size of any hoops or crop covers you plan to use when designing your beds. Additionally, it's crucial to plan for irrigation before forming beds and commencing planting. Further details on irrigation will be discussed in a subsequent post.

For smaller gardens, marking the bed width with a 36-inch board and using a hoe or hand tool to dig a furrow along the length of the bed, typically 8-10 inches deep, can suffice. However, for larger areas, investing in a tractor attachment can expedite the process by assisting in cutting, leveling, and shaping the beds.

Cultivating a farm involves a systematic approach to prepare the soil, plant crops, manage weed growth, and maintain overall soil health. The process begins with assessing soil quality, pH level, moisture content, and nutrient composition, while also considering the specific requirements of the crops to be grown in terms of soil type, drainage, and sunlight.

Once the soil assessment is complete, the next step is soil preparation. This entails tilling or plowing the land to loosen it and create an optimal seedbed. The depth and method of tillage depend on factors such as soil type, crop type, and preferred farming practices, whether conventional or no-till.

Based on soil test results and crop requirements, organic matter such as compost or manure, along with soil amendments like lime or gypsum, are added to improve soil structure, fertility, and nutrient availability. These amendments are incorporated into the soil through tilling or mixing.

After soil preparation and amendment, crops are planted according to recommended spacing, depth, and timing. Considerations include seed quality, planting equipment, and weather conditions to ensure optimal seed germination and establishment.

Managing weed growth is crucial to minimize competition for nutrients, water, and sunlight. Weed control measures may include mechanical methods such as cultivation or hoeing, mulching, or selective and judicious herbicide application.

Figure 34: Row crop cultivation. Einboeck.official, CC BY-SA 4.0, via Wikimedia Commons.

Proper irrigation and water management are essential to maintain adequate moisture levels for crop growth. Monitoring soil moisture regularly and adjusting irrigation schedules prevent under- or over-watering.

Fertilizers are applied as needed to supplement soil nutrients and support crop growth. Soil test results guide the type, amount, and timing of fertilizer applications, aiming for balanced nutrient management to maximize yields and quality.

Monitoring crops for pests, diseases, and other threats to plant health is necessary. Integrated pest management strategies combine cultural, biological, and chemical control methods to minimize crop damage and losses.

Regular field inspections ensure crop health, and agronomic practices like pruning, thinning, and staking optimize plant growth and yield potential.

Harvesting crops at the appropriate maturity stage with suitable equipment minimizes damage and preserves quality. Proper post-harvest handling, storage, and transportation maintain crop freshness and marketability.

Implementing soil conservation practices such as crop rotation, cover cropping, and minimal tillage prevent erosion, improve soil structure, and enhance long-term productivity and sustainability.

Accurate record-keeping of farm activities, inputs, and yields facilitates performance tracking, trend identification, and informed decision-making for future planning and management adjustments.

Adapting these steps to the farm's specific conditions and requirements ensures effective land cultivation, successful crop production, and the promotion of soil health and environmental sustainability.

Plowing with Two Row Cultivators:

1. Begin by planting your garden with the rows spaced appropriately.
    - Ensure that the spacing between rows is suitable for the available equipment.
    - For small vegetable patches, consider using a tiller or hoe for efficiency.
2. Start early in the day to take advantage of cooler temperatures and optimal moisture levels for cultivation.
    - Moisture is crucial for effective cultivation, and morning hours often offer better conditions.
3. Set the plows to the desired spacing to maintain moisture near the plants' roots while preventing weed growth in the furrows.
    - Adjust the spacing of the plows to accommodate the size and spreading root systems of your plants.
4. Verify that the plow points are set at the correct depth and

positioned appropriately on the cultivator toolbar.

- Ensure each plow complements adjacent ones for efficient operation.

5. Adjust the pitch of the plows to control how much soil is thrown into the planting furrows.
    - Forward pitch reduces soil accumulation, while backward pitch increases it.
6. Properly space the tractor tyres to avoid damaging plant roots during plowing.
    - Adjust the tyre width to match the row spacing, preventing interference with plant roots.
7. Adjust sway chains or bars to allow free movement for the plows along the rows.
    - Ensure the plows can follow the row path without hindrance from the cultivator's frame.
8. Set the plow depth according to soil type and desired tillage level.
    - Loamy soil may require deeper plowing for optimal root development, while sandy soil may need shallower plowing to retain moisture.
9. Choose appropriate timing for plowing based on plant size and weed growth.
    - Plow when plants are large enough to avoid covering them with displaced soil but while weeds are still small enough to be effectively controlled.

10. Ensure soil moisture levels are sufficient before plowing to maximize the hilling effect and moisture retention.
    - Plowing in dry weather may lead to rapid moisture loss, while plowing just before rainfall may reduce weed control effectiveness.
11. Maintain a consistent speed during plowing to achieve desired results and minimize soil disturbance.
    - Faster speeds may result in more soil being thrown into the furrows.
12. Position tractor wheels between rows using guide points to maintain proper spacing.
    - Guide points on the tractor frame help ensure the correct alignment between furrows.
13. Exercise caution while steering to avoid damaging crops.
    - Careful steering is essential, especially in curved or irregularly planted rows.
14. Periodically remove weeds and debris from plow points to prevent soil buildup and ensure optimal operation.
    - Accumulated debris can lead to soil overflow, covering crop plants during plowing.

Completing Cultivator Operations

Conducting shutdown procedures for cultivators in accordance with workplace guidelines is essential for safety and equipment maintenance. This involves several key steps, including turning off the engine and auxiliary systems, engaging safety mechanisms, and securely stowing attachments. It's crucial to park the cultivator on level ground in a designated area and complete any required paperwork or checklists to document the shutdown process. Additionally, communicating with supervisors or colleagues to confirm the cultivator is safely shut down and secured ensures compliance with workplace procedures.

Regular servicing and maintenance are vital for keeping cultivators in optimal working condition. Following the manufacturer's maintenance schedule and guidelines is essential. This includes inspecting the cultivator for signs of wear, damage, or leaks, checking fluid levels, lubricating moving parts, and cleaning filters, screens, and air intakes. Addressing any minor issues promptly helps prevent them from becoming major problems, ensuring continued productivity and efficiency.

Identifying and reporting malfunctions, faults, irregular performance, or damage is crucial for maintaining safety and productivity on the farm. Monitoring the cultivator during operation for any unusual sounds, vibrations, or performance issues is essential. If any signs of malfunction or irregular performance are noticed, operations should be stopped immediately, and the issue reported to supervisors or maintenance personnel. Providing detailed information about the observed issue helps facilitate timely repairs or adjustments by trained personnel.

Proper cleaning, storage, and security measures are necessary to prolong the lifespan of cultivating machinery and equipment. After each use, the cultivator should be thoroughly cleaned to remove dirt, debris, and plant residues. It should then be stored in a designated area protected from weather exposure and potential damage. Securing the cultivator with locking mechanisms or other security measures prevents unauthorized access or tampering, ensuring its safety and integrity.

Safely storing cultivator keys is essential for preventing unauthorized use and ensuring security. This involves turning off the cultivator and removing the keys from the ignition, then storing them in a secure location designated by the workplace, such as a lockbox or key cabinet. Keeping the keys out of reach of unauthorized individuals and following any additional key management protocols specified by the organization further enhances security measures.

Maintaining accurate records of cultivator usage is essential for tracking maintenance, performance, and operational history. This involves recording key information such as operating hours, maintenance tasks performed, repairs conducted, and any incidents or issues encountered during operation. Using the required format and documentation tools specified by the workplace ensures consistency and compliance. Making records accessible to relevant personnel, such as supervisors, maintenance staff, or safety inspectors, facilitates compliance and decision-making processes.

Chapter Four

Windrowers/ Swathers

A windrower, also known as a swather or hay swather, is a type of agricultural machinery used primarily in hay and forage harvesting operations. It is designed to cut and swath crops such as grass, alfalfa, clover, or other forage crops, laying them in neat rows or windrows for subsequent drying or collection.

Its key features and uses include:

1. Cutting: The windrower is equipped with a cutting mechanism, typically a sickle bar or a rotary cutter, mounted on the front of the machine. This cutting mechanism slices through standing crops at a consistent height, severing the stems and leaves cleanly.

2. Swathing: After cutting, the windrower places the crop material in rows, known as windrows, on the ground behind the machine. These windrows are typically formed in a uniform manner to facilitate drying or further processing.

3. Drying: Windrowing allows forage crops to dry more efficiently by exposing a larger surface area to air and sunlight. This process is crucial for haymaking, as proper drying helps preserve the

nutritional quality of the forage and prevents mold or spoilage during storage.

4. Preparation for Baling or Harvesting: In many cases, the windrowed crop is left in the field to dry before being baled for hay or collected for silage. Windrowing facilitates subsequent harvesting operations by organizing the crop material into manageable rows, making it easier to gather and process.

5. Regulation of Crop Height: Windrowers often feature adjustable cutting heights, allowing operators to control the length of the crop material left in the field. This flexibility enables farmers to optimize cutting heights based on crop maturity, weather conditions, and intended uses of the harvested forage.

6. Efficiency and Productivity: By combining cutting and windrowing functions into a single operation, windrowers help improve the efficiency and productivity of forage harvesting. They allow farmers to complete harvesting tasks more quickly and effectively, especially during critical periods when weather conditions are favourable.

Overall, windrowers play a vital role in modern forage production systems, enabling farmers to efficiently harvest, process, and preserve high-quality forage crops for use as livestock feed or other agricultural purposes. Windrowers, employed by farmers, serve the purpose of cutting, segregating, and forming rows of hay or small grain crops. These machines are generally self-propelled and employ a sickle bar to perform the cutting action on the crops. Additionally, they utilize a reel mechanism to lift the crops onto a conveyor belt. Subsequently, the conveyor belt transfers the crops into what is known as a windrow, characterized by its elongated and narrow formation of cut crops.

Figure 35: A self-propelled swather by New Holland. Kowloonese, CC BY 3.0 , via Wikimedia Commons.

A swather as known in the US, known as a windrower in Australia and elsewhere, can either be self-propelled, featuring an engine, or towed by a tractor and powered by a power take-off mechanism. Utilizing a reciprocating sickle bar or rotating discs, a swather severs the stems of the crops. A reel assists in neatly guiding the cut crops onto a canvas or auger conveyor, which then deposits them into a windrow, keeping the stems aligned and supported above the ground by the stubble.

Functionally, a swather accomplishes the same tasks for hay crops as manual methods like hand scything, cradling and swathing, or mowing and raking. Additionally, horizontal rollers situated behind the cutters may be employed to crimp or condition the stems of hay crops, reducing their drying time.

In the context of grain crops, as combines gradually replaced threshing machines, swathers introduced an optional step in the harvesting process to allow for the necessary drying time, previously facilitated by binding. While swathing remains more prevalent in the northern United States and Canada, regions with longer growing seasons often opt for direct harvesting of standing grain crops by combines. However, advancements in crop varieties capable of rapid maturation have diminished the need for swathing grains even in northern regions.

Apart from expediting the drying of ripe grain, windrowing the entire growing crop ensures uniform ripening and dehydration of stalks and green weeds, aiding in effective post-threshing winnowing and separation of grain and other material. Alternatively, chemical desiccation using substances like glyphosate, paraquat, or diquat has been employed to enable direct combining of weedy or unevenly ripened standing crops.

In the northern United States and Canada, swathing is predominantly employed due to the shorter growing seasons, allowing for the reduction of curing time for grain by cutting the plant stems. Conversely, in regions with longer growing seasons, hay and cereal crops are often left standing and harvested using a combine. This is because standing grains in areas with extended growing seasons can usually attain the necessary moisture level without additional intervention.

Windrowers, also known as swathers, come in various types, each tailored to specific agricultural needs. Here are the different types of windrowers commonly used:

1. Self-Propelled Windrowers:

    - Self-propelled windrowers are equipped with their own engines, allowing them to operate independently without being towed by a tractor.
    - These windrowers are highly manoeuvrable and efficient, ideal for large-scale operations where speed and productivity are crucial.

  - They often feature cutting mechanisms such as sickle bars or rotating discs to sever crop stems and form windrows.

2. Tractor-Drawn Windrowers:
   - Tractor-drawn windrowers are pulled by a tractor and powered by its power take-off (PTO) mechanism.
   - They are more common in smaller farms or where the use of a dedicated self-propelled windrower is not economically feasible.
   - These windrowers typically have similar cutting mechanisms as self-propelled ones, such as sickle bars or rotating discs.
3. Reciprocating Sickle Bar Windrowers:
   - These windrowers use a reciprocating sickle bar to cut crop stems.
   - The sickle bar moves back and forth in a horizontal motion, cutting the crop as it moves along.
   - They are suitable for cutting hay or small grain crops and forming windrows for drying.
4. Rotary Disc Windrowers:
   - Rotary disc windrowers utilize rotating discs to sever crop stems.
   - The discs rotate rapidly, slicing through the crop as they move forward, creating windrows in the process.
   - They offer faster cutting speeds and are often preferred for larger-scale operations or when dealing with tougher crop

varieties.

5. Conditioning Windrowers:

    - Some windrowers come equipped with conditioning rollers or crimpers behind the cutting mechanism.
    - These rollers or crimpers help crush or crimp the crop stems, aiding in the drying process by increasing the surface area exposed to air.
    - Conditioning windrowers are commonly used for hay crops to accelerate drying and improve overall hay quality.

Overall, the choice of windrower type depends on factors such as farm size, crop type, terrain, and budget. Each type offers unique advantages and capabilities to meet the diverse needs of modern agricultural operations.

Figure 36: Tractor drawn windrower. Windrower at Forthill Farm by Oliver Dixon, CC BY-SA 2.0 , via Wikimedia Commons.

Windrowers typically consist of several key components that work together to efficiently harvest crops. Here's a detailed explanation of how a windrower works:

1. Cutting Mechanism:

   - The cutting mechanism is the primary component responsible for cutting the crop. Windrowers commonly use either a reciprocating sickle bar or rotating discs for this purpose.

   - Reciprocating Sickle Bar: In windrowers equipped with a sickle bar, the bar moves back and forth in a horizontal motion, cutting the crop stems as it advances.

   - Rotary Discs: Alternatively, some windrowers utilize rotating discs that spin rapidly to slice through the crop as they move forward.

2. Reel:

    - The reel is a component located in front of the cutting mechanism. It consists of rotating rods or fingers that help guide the standing crop into the cutting mechanism.

    - The reel ensures that the crop is uniformly fed into the cutting mechanism, improving cutting efficiency and reducing crop loss.

3. Conveyor Belt or Auger:

    - After the crop is cut by the sickle bar or discs, it is lifted and carried by a conveyor belt or auger.

    - The conveyor belt or auger transports the cut crop to the centre of the windrower, where it is deposited into a windrow.

4. Formation of Windrow:

    - The cut crop is deposited onto the ground in a long, narrow row known as a windrow. The width of the windrow can be adjusted based on the specific requirements of the operation.

    - As the windrower moves forward, it continues to cut and deposit the crop, forming successive windrows in the field.

5. Conditioning (Optional):

    - Some windrowers are equipped with conditioning rollers or crimpers located behind the cutting mechanism.

    - These rollers or crimpers help crush or crimp the crop stems, which can accelerate the drying process by increasing the surface area exposed to air.

- Conditioning windrowers are commonly used for hay crops to improve drying efficiency and overall hay quality.

6. Adjustments and Control:
   - Windrowers typically feature adjustable settings that allow operators to control various aspects of the harvesting process, such as cutting height, reel speed, and windrow width.
   - Operators may need to make adjustments based on factors such as crop type, field conditions, and desired drying time.
7. Harvesting Operation:
   - During operation, the windrower is driven across the field in a systematic pattern, cutting and forming windrows as it progresses.
   - The harvested crop can then be left to dry in the windrows before being further processed or collected for storage.

In summary, a windrower works by cutting hay or small grain crops using a cutting mechanism, guiding the cut crop into a conveyor system, and depositing it into windrows on the ground. It offers efficient and effective harvesting capabilities for agricultural operations, helping to prepare crops for subsequent processing or storage.

The optimal width of a windrower is determined by the anticipated yield of the crop and the dimensions of the harvester. Wider windrowers are preferable for crops with lower yields and shorter varieties, as they minimize material passing through the machine. For instance, tractor-drawn windrowers with a width of 10 meters are suitable for crop yields ranging from 1.5 to 2 tons per hectare. Alternatively, MacDon draper fronts, measuring 12 meters wide and attached to the header, perform well for crops with expected yields below 1.5 tons per hectare. Larger windrows may take slightly longer to achieve the desired harvest moisture content of eight percent. However, smaller or lighter windrows are susceptible to being blown by the wind or lifted by whirlwinds, posing challenges during harvest. Typically, a well-sized windrow is approximately 1.5 meters wide and 1 meter high. The positioning of the windrow on the stubble is crucial as it affects air circulation underneath, aiding in drying. While most large-capacity harvesters can handle sizable windrows, modifications may be necessary in high-yielding areas (>2.5 tons per hectare), such as raising the height and switching to a larger diameter table auger. Dealers are best suited to set up these modifications. Matching the harvesting capacity of the header with the size of the windrow is essential; smaller capacity harvesters generally require smaller windrows. Large headers, particularly those equipped with hydrostatic drives and axial flow mechanisms, can effectively manage large windrows when equipped with appropriately sized pickup or draper fronts.

Several hazards are associated with windrowers/swathers, including:

1. Entanglement: Windrowers have moving parts such as reels, belts, augers, and other components that can catch loose clothing, hair, or body parts, leading to entanglement injuries.

2. Falls: Operators may fall from the platform or step when mounting or dismounting the machine, especially if it's in motion or on uneven terrain.

3. Collision: Windrowers operate in fields alongside other machinery, vehicles, or obstacles. Collisions with stationary objects or other moving equipment can occur, leading to injuries or damage.

4. Pinch points: Moving parts of the windrower machinery create pinch points where body parts or clothing can become trapped, causing crush injuries or amputations.

5. Electrocution: Electrical components within the windrower can pose a risk of electrocution if damaged or improperly handled, especially in wet conditions.

6. Fire hazard: Accumulation of dry vegetation or crop residue around hot engine components or exhaust systems can lead to fires, particularly during dry conditions or in the presence of flammable materials.

7. Chemical exposure: Windrowers may be used in conjunction with chemical applications for weed control or crop management. Exposure to pesticides or herbicides can occur through contact with the skin, inhalation, or ingestion.

8. Noise: Windrowers generate significant noise levels during operation, which can lead to hearing damage or impairment if proper hearing protection is not worn by operators.

9. Rollover: Windrowers have a high centre of gravity, especially when loaded with crop material, making them prone to tipping over on uneven terrain or steep slopes, potentially causing serious injuries or fatalities to operators.

10. Machinery malfunctions: Mechanical failures or malfunctions of components such as brakes, steering systems, or hydraulic systems can lead to loss of control and accidents.

To mitigate these hazards, proper training, maintenance, and adherence to safety protocols are essential. Operators should be trained in safe operation procedures, wear appropriate personal protective equipment, conduct pre-operation checks, and follow manufacturer guidelines and regulatory requirements. Regular maintenance and inspection of machinery are also crucial to ensure safe operation and prevent accidents.

However, using windrowers introduces further safety considerations. One such consideration is the use of auger or canvas draper mechanisms to transport hay to the centre of the platform. Auger platforms, in particular, pose potential safety hazards due to their status as additional moving parts and their susceptibility to getting plugged up by crops. When unplugging such mechanisms, operators must exercise caution to avoid the hazards of shearing and pinching actions. Standard precautions observed with mower-conditioners, including disengaging the power take-off (PTO) and shutting off the engine before performing any maintenance or repair work on the platform, should also be followed with windrowers.

Another safety consideration with windrowers is their speed, which is influenced by terrain conditions and crop density. Operating over rough terrain or on hillsides requires careful attention to avoid hazards such as holes or obstacles that could cause the windrower to tip over or throw the operator from the machine. Additionally, high operating speeds in heavy crops can lead to frequent plugging, increasing the risk of accidents. Adjusting the speed according to the terrain and crop density not only enhances safety but also improves efficiency by reducing the need for frequent stops to clear the machine.

Windrowers feature steering mechanisms comprising steering levers and a steering wheel. Operators utilize steering levers for sharp turns at the end of the field and adjust speed accordingly to navigate safely. When making turns on hillsides, reducing speed and exercising caution are essential to prevent overturning.

The height at which the platform is operated varies depending on the crop being harvested. While harvesting hay crops, the platform may operate very close to the ground. Operators must remain vigilant for rocks or obstacles that could be struck or picked up by the platform, as well as irregularities in the ground that pose safety hazards.

Planning and Preparing for Windrowing

To operate the platform safely, it's crucial to adhere to several guidelines: Ensure that all shields are securely in place to protect against potential hazards. Avoid manually feeding material into the machine, as this poses a significant risk of injury. Refrain from leaning against, sitting on, or standing on the cutterbar curtains or their supporting framework to prevent accidents. Never operate the platform with the cutterbar in a raised position to maintain stability and prevent damage. Regularly inspect and tighten disks and knife bolts to ensure proper functionality. For machines equipped with an impeller conditioner, verify that the tines are securely fastened on the rotor. Always operate the machine at the rated speed specified by the manufacturer and drive slowly when traversing rough terrain to maintain control and stability.

Prior to commencing operation, take several preparatory steps: Familiarize yourself with the operator's manual, machine decals, and the safety section of the manual to understand proper procedures and precautions. Remove any foreign objects from the machine that could interfere with its operation. Take the time to acquaint yourself with all controls that affect the machine's functions to operate it safely and efficiently. Ensure that no one is in the vicinity of the machine before starting it, and never allow anyone to ride on or near the machine while it is running to prevent accidents. Before starting the machine, confirm that all shields and guards are in place and in good condition to provide adequate protection during operation. These precautions are essential for ensuring the safe and effective operation of the platform.

It is imperative to prohibit riders from being on the machine during operation due to the inherent risks involved. Riders are vulnerable to potential injuries such as being struck by foreign objects or being thrown off the machine into the platform or under the wheels, which can result in severe harm. Anyone permitted to ride on the machine must be seated in the designated training seat with the seat belt securely fastened. Additionally, the presence of riders obstructs the operator's view, leading to unsafe operation of the machine.

The operating, steering, and braking performance of the windrower can be significantly impacted by the size of the platform used, which alters the centre of gravity of the machine. To ensure proper ground contact and stability, it is necessary to add ballast to the rear of the windrower as recommended for the specific platform in use.

Many moving parts of the windrower, such as the cutterbar and conditioner rolls, cannot be fully shielded due to their function. Therefore, it is essential to avoid contact with these moving parts during operation to prevent accidents. Before servicing or manually unplugging the platform, follow a series of safety steps to ensure the machine is securely parked and powered off.

Entanglement in a rotating driveline can result in serious injury or even death, emphasizing the importance of maintaining caution around these areas. Proper shielding covering the conditioner should always be in place, and individuals should wear close-fitting clothing to minimize risks. Additionally, the engine must be stopped, keys removed, and the driveline fully stopped before performing any maintenance or adjustments.

Careful attention must be given to prevent injuries from thrown objects, as rotating cutting blades can propel stones and other debris significant distances. Utilizing cutterbar curtains is crucial to reducing the risk of thrown objects, and these curtains should always be kept down during platform operation. It is essential to replace worn or damaged curtains promptly.

Safe handling of fuel is paramount to avoid fire hazards, necessitating precautions such as avoiding refuelling near open flames or sparks and stopping the engine before refuelling. Proper maintenance and cleanliness of the machine are also vital to prevent fires caused by accumulated debris or spilled fuel.

Being prepared for emergencies is essential for swift and effective responses to potential hazards or accidents. Keeping a first aid kit and fire extinguisher readily accessible and maintaining a list of emergency contact numbers can help mitigate risks and ensure prompt assistance when needed.

Starting fluid poses fire risks due to its flammability, necessitating careful handling to prevent accidents. Proper storage, avoiding sparks or flames during use, and refraining from incinerating or puncturing containers are crucial safety measures to minimize risks associated with starting fluid.

Wearing appropriate protective clothing and safety equipment is essential to minimize the risk of injuries or health issues while operating the machine. This includes wearing close-fitting clothing and using hearing protection to prevent hearing impairment from prolonged exposure to loud noises.

Utilizing safety lights and devices is vital to enhance visibility and prevent collisions, especially when operating the machine on public roads or in areas with other road users. Following local regulations for equipment lighting and marking and maintaining visibility of lights and markings are essential safety practices.

Proper procedures must be followed when transporting the windrower with the platform attached to ensure stability and safety. This includes raising the platform, engaging the platform lift lockout lever, and thoroughly checking the surrounding area for bystanders or obstructions before driving away.

A self-propelled windrower typically features a set of operator controls designed to facilitate safe and efficient operation. These controls allow the operator to manoeuvre the machine, adjust various settings, and engage or disengage specific functions as needed. Here are some typical operator controls found in a self-propelled windrower:

1. Steering Wheel: The steering wheel enables the operator to control the direction of the windrower. By turning the steering wheel left or right, the operator can steer the machine accordingly.

2. Hydrostatic Drive Lever: This lever controls the speed and direction of the windrower's movement. By adjusting the position of the hydrostatic drive lever, the operator can move the windrower forward or backward and control its speed.

3. Platform Drive Control: The platform drive control engages or disengages the drive mechanism responsible for powering the cutting platform or header. This control allows the operator to start or stop the cutting mechanism as needed.

4. Platform Height Adjustment: Some windrowers feature controls for adjusting the height of the cutting platform. These controls allow the operator to raise or lower the platform to accommodate varying crop conditions or terrain.

5. Throttle Control: The throttle control regulates the engine speed, allowing the operator to increase or decrease the power output of the windrower's engine.

6. Parking Brake: The parking brake is used to secure the windrower in place when it is parked or not in use. Engaging

the parking brake prevents the windrower from moving unintentionally.

7. Emergency Stop Button: An emergency stop button or switch is often provided for quickly shutting down the machine in case of an emergency or hazardous situation.

8. Indicator Lights and Gauges: Indicator lights and gauges provide essential information to the operator regarding the status of various systems, such as engine temperature, fuel level, hydraulic pressure, and battery voltage.

9. Safety Interlock System: Many windrowers are equipped with a safety interlock system that prevents certain functions from being activated unless specific conditions are met. For example, the cutting platform may only operate when the windrower is moving at a certain speed or when the parking brake is disengaged.

10. Accessory Controls: Depending on the model and configuration, a self-propelled windrower may feature additional controls for operating optional accessories or attachments, such as lights, radios, or GPS systems.

Overall, these operator controls play a crucial role in enabling the operator to effectively operate and control a self-propelled windrower while ensuring safety and productivity during harvesting operations.

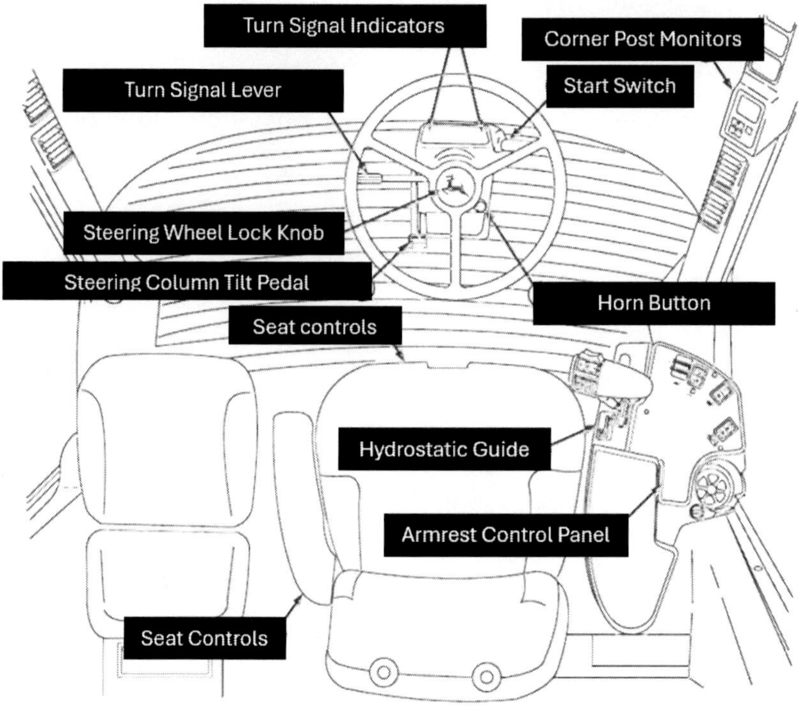

Figure 37: Sample operator cabin layout and controls.

The armrest control panel may consist of various components, including platform control switches for managing the cutting platform's operation, a hydrostatic drive control lever to adjust the windrower's speed and direction, a platform drive engage switch for activating or deactivating the platform's drive mechanism, an engine speed adjustment knob for regulating the engine's power output, a platform float pressure adjustment feature, a field/road speed switch to toggle between different speed settings, a beverage holder for convenience during operation, and a power port for charging or connecting electronic devices.

Prior to commencing operations with a self-propelling windrower, several prestart checks need to be performed to ensure the equipment's safety and functionality. These checks typically include inspecting the overall condition of the machine, such as examining for any visible signs of damage, leaks, or wear and tear. Additionally, it's essential to check the fluid levels, including fuel, engine oil, hydraulic fluid, and coolant, and refill them if necessary to ensure optimal performance. Verifying the functionality of critical components, such as brakes, steering, lights, and warning signals, is crucial to ensure safe operation. Furthermore, inspecting the cutting platform and conditioning components for any obstructions, damage, or signs of wear is essential for efficient and safe operation during harvesting. Finally, it's imperative to follow the manufacturer's guidelines outlined in the operator's manual for specific prestart checks tailored to the windrower model to ensure compliance with safety standards and regulations.

Prior to windrowing, several planning and preparation steps are necessary to ensure a smooth and efficient operation:

1. Crop Assessment: Assess the maturity and condition of the crop to determine if it's ready for windrowing. Ensure the crop has reached the appropriate moisture content for optimal drying during windrowing.

2. Equipment Inspection: Thoroughly inspect the windrower and associated equipment, including the cutting platform, conditioning components, and driveline, to ensure they are in good working condition. Check for any signs of damage, wear, or malfunction and address any issues promptly.

3. Adjustments: Make necessary adjustments to the windrower settings based on the crop type, density, and field conditions. Adjust the cutting height, reel speed, and conditioning settings to achieve the desired windrow formation and crop conditioning.

4. Field Preparation: Prepare the field by removing any obstacles or debris that could interfere with the windrowing operation. Clear the area of rocks, branches, and other debris to prevent damage to the equipment and ensure a smooth harvesting

process.

5. Weather Monitoring: Monitor weather forecasts closely to choose the optimal time for windrowing. Aim for dry and sunny weather conditions to facilitate rapid drying of the windrowed crop and minimize the risk of spoilage.

6. Safety Measures: Implement appropriate safety measures to ensure the well-being of operators and bystanders during windrowing operations. Provide proper training to operators on safe equipment operation and emergency procedures. Additionally, ensure all safety guards and shields are in place and functioning correctly.

7. Communication: Coordinate with other personnel involved in the harvesting process, such as truck drivers and field supervisors, to ensure efficient coordination and timely transport of windrowed crops to storage facilities.

Windrowing Operations

Swathing, also known as windrowing, is a technique used in agricultural harvesting to cut and arrange crops into rows, facilitating drying and subsequent collection. Different methods of windrowing are employed based on factors such as crop type, field conditions, and equipment capabilities. The following is an overview of various windrowing techniques:

1. Swathing (Single Windrowing): In this method, the crop is cut by a swather or windrower and deposited into a single row or windrow on the field. Swathing is commonly used for crops like grains, oilseeds, and forages. The single windrow allows

for efficient drying and facilitates subsequent harvesting with machinery like combines.

2. Single Windrowing: This technique involves cutting the crop and depositing it into a single windrow. It's similar to swathing but may refer to the process when done with different equipment or methods. Single windrowing is suitable for crops where a single row is sufficient for drying and harvesting.

3. Double Windrowing: In double windrowing, the crop is cut and deposited into two parallel windrows side by side. This method is often used for crops with high yield or when conditions require faster drying. Double windrowing increases drying efficiency by spreading the crop thinner across two rows.

4. Side-By-Side Double Windrowing: Similar to double windrowing, this method involves creating two windrows, but the rows are positioned closer together, side by side. Side-by-side double windrowing is useful in situations where space is limited or when the windrower needs to cover the entire width of the field efficiently.

5. Two-In-One Double Windrowing: In this technique, two windrows are formed side by side, but each windrow contains two smaller sub-windrows. The two-in-one double windrowing method is often used for crops with high yield or when efficient drying is crucial. It allows for increased surface area exposure to air, promoting faster and more uniform drying.

Each windrowing method has its advantages and is chosen based on factors such as crop type, field conditions, equipment availability, and desired drying efficiency. Proper selection and execution of windrowing techniques contribute to optimal harvesting and crop quality.

Figure 38: Hay windrows. Rhian, CC BY 2.0, via Wikimedia Commons.

Swathing, also known as single windrowing, is a method commonly used in agricultural practices, particularly for crops like grains, oilseeds, and forages. This process involves cutting the crop using a swather or windrower and depositing it into a single row or windrow on the field. Swathing serves several purposes, including facilitating drying, optimizing harvesting efficiency, and preserving crop quality.

The process of swathing typically involves several key steps:

1. Preparation and Planning: Before initiating swathing, farmers must assess the crop's maturity and condition to determine the optimal timing for harvest. This includes monitoring factors such as moisture content, plant health, and weather forecasts. Proper planning ensures that swathing is performed at the right time to maximize crop yield and quality.

2. Equipment Setup: Farmers prepare the swather or windrower

for operation by adjusting settings such as cutting height, reel speed, and conditioner settings based on the crop type and field conditions. Proper equipment setup ensures efficient cutting and uniform windrow formation.

3. Field Operation: Once the equipment is set up, operators begin swathing by driving the swather or windrower across the field in a systematic pattern. The cutting mechanism of the equipment slices through the standing crop, severing the stems at the desired height. As the crop is cut, it is deposited into a single row or windrow on the ground behind the equipment.

4. Windrow Formation: The swathed crop forms a single windrow along the length of the field, typically oriented in the direction of prevailing winds to facilitate drying. The width and density of the windrow can be adjusted based on crop type, field conditions, and harvesting preferences. Proper windrow formation is essential for efficient drying and subsequent harvesting operations.

5. Drying and Conditioning: After swathing, the crop undergoes a drying process facilitated by exposure to sunlight, air circulation, and natural evaporation. Depending on environmental conditions, crop moisture content, and weather forecasts, farmers may use conditioning equipment such as rollers or conditioners to further accelerate drying and improve crop quality.

6. Monitoring and Management: Throughout the swathing process, farmers monitor the drying progress and field conditions to ensure optimal outcomes. This includes assessing crop moisture levels, checking for signs of weather-related damage or spoilage, and making adjustments to equipment or harvesting plans as needed.

7. Subsequent Harvesting: Once the swathed crop has reached the desired moisture level and is sufficiently dried, farmers proceed with the subsequent harvesting phase using machinery such as combines or forage harvesters. The single windrow configuration facilitates efficient harvesting by allowing the equipment to gather and process the crop with minimal losses and maximum throughput.

Double windrowing involves cutting the crop and depositing it into two parallel windrows side by side. Double windrowing is preferred in situations where single windrows may not allow for efficient drying or when there is a need to increase the throughput of harvesting operations.

The process of double windrowing includes:

1. Preparation and Planning: As with any harvesting operation, thorough preparation and planning are essential for successful double windrowing. Farmers assess the crop's maturity, moisture content, and overall condition to determine the optimal timing for harvest. Factors such as weather forecasts, field conditions, and equipment availability are also taken into consideration during the planning phase.

2. Equipment Setup: Before initiating double windrowing, farmers ensure that the swather or windrower is properly set up and calibrated for the specific crop and field conditions. This includes adjusting cutting heights, reel speeds, and conditioner settings to achieve uniform windrow formation and maximize drying efficiency. Additionally, operators may adjust the equipment's width to accommodate the desired spacing between the two windrows.

3. Field Operation: Once the equipment is properly set up, operators drive the swather or windrower across the field in a

systematic pattern, cutting the crop and depositing it into two parallel windrows. The cutting mechanism of the equipment slices through the standing crop, creating two rows of cut vegetation on the ground. Operators maintain a consistent speed and trajectory to ensure uniform windrow formation and coverage across the field.

4. Windrow Formation: As the crop is cut and deposited, it forms two parallel windrows side by side. The spacing between the windrows can be adjusted based on crop type, field conditions, and harvesting preferences. Proper windrow formation is critical for efficient drying and subsequent harvesting operations. The windrows should be oriented in the direction of prevailing winds to promote air circulation and facilitate drying.

5. Drying and Conditioning: After double windrowing, the crop undergoes a drying process facilitated by exposure to sunlight, air circulation, and natural evaporation. The thinner spread of the crop across two windrows allows for faster drying compared to single windrowing. Farmers may monitor moisture levels and weather conditions to optimize drying efficiency. Conditioning equipment such as rollers or conditioners may be used to further accelerate drying if necessary.

6. Monitoring and Management: Throughout the windrowing process, farmers monitor the drying progress and field conditions to ensure optimal outcomes. This includes regular assessments of crop moisture levels, windrow integrity, and weather forecasts. Farmers may make adjustments to equipment settings or harvesting plans as needed to maximize drying efficiency and crop quality.

7. Subsequent Harvesting: Once the crop has reached the de-

sired moisture level and is sufficiently dried, farmers proceed with subsequent harvesting operations using machinery such as combines or forage harvesters. The parallel windrows created by double windrowing facilitate efficient harvesting by allowing equipment to gather and process the crop with minimal losses and maximum throughput.

Side-by-side double windrowing is a technique used in agricultural harvesting, particularly when space is limited or when the windrower needs to efficiently cover the entire width of the field. Similar to traditional double windrowing, this method involves creating two windrows, but the rows are positioned closer together, side by side. This allows for increased efficiency and flexibility in field operations.

To perform side-by-side double windrowing:

1. Preparation and Planning: As with any harvesting operation, thorough preparation and planning are essential. Farmers assess the field conditions, crop maturity, and weather forecasts to determine the optimal timing for harvest. They also ensure that the windrower is properly set up and calibrated for the specific crop and field conditions.

2. Equipment Setup: Before initiating side-by-side double windrowing, farmers ensure that the windrower is set up correctly. This includes adjusting the cutting height, reel speed, and conditioner settings to achieve uniform windrow formation. Additionally, operators may adjust the equipment's width to accommodate the desired spacing between the two windrows.

3. Field Operation: Once the equipment is properly set up, operators drive the windrower across the field in a systematic pattern. As the windrower moves forward, it cuts the crop and deposits it into two windrows positioned side by side. Operators maintain a steady speed and trajectory to ensure uniform windrow

formation and coverage across the field.

4. Windrow Formation: As the crop is cut and deposited, it forms two parallel windrows positioned closely together. The spacing between the windrows is adjusted based on field conditions and equipment specifications. Proper windrow formation is essential for efficient drying and subsequent harvesting operations.

5. Drying and Conditioning: After side-by-side double windrowing, the crop undergoes a drying process facilitated by exposure to sunlight, air circulation, and natural evaporation. The close proximity of the windrows may enhance drying efficiency by maximizing air circulation and reducing drying time. Farmers may monitor moisture levels and weather conditions to optimize drying efficiency.

6. Monitoring and Management: Throughout the windrowing process, farmers monitor the drying progress and field conditions to ensure optimal outcomes. This includes regular assessments of crop moisture levels, windrow integrity, and weather forecasts. Farmers may make adjustments to equipment settings or harvesting plans as needed to maximize drying efficiency and crop quality.

7. Subsequent Harvesting: Once the crop has reached the desired moisture level and is sufficiently dried, farmers proceed with subsequent harvesting operations using machinery such as combines or forage harvesters. The closely positioned windrows created by side-by-side double windrowing facilitate efficient harvesting by allowing equipment to gather and process the crop with minimal losses and maximum throughput.

Two-in-one double windrowing is an advanced technique used in agricultural harvesting, particularly for crops with high yield or when efficient drying is crucial. In this method, two windrows are formed side by side, but each windrow consists of two smaller sub-windrows. This approach enhances drying efficiency by increasing the surface area exposure to air, thereby promoting faster and more uniform drying of the crop.

Here's a detailed explanation of how to perform two-in-one double windrowing:

1. Preparation and Planning: Before initiating the two-in-one double windrowing process, farmers assess the field conditions, crop maturity, and weather forecasts to determine the optimal timing for harvest. They also ensure that the windrower is properly set up and calibrated for the specific crop and field conditions.

2. Equipment Setup: Farmers adjust the windrower's settings to accommodate the two-in-one double windrowing technique. This may involve configuring the equipment to form two smaller sub-windrows within each main windrow. Operators adjust the cutting height, reel speed, and conditioner settings to achieve uniform windrow formation and maximize drying efficiency.

3. Field Operation: Once the equipment is properly set up, operators drive the windrower across the field in a systematic pattern. As the windrower moves forward, it cuts the crop and deposits it into two windrows positioned side by side. However, instead of forming single windrows, the equipment creates two smaller sub-windrows within each main windrow.

4. Sub-Windrow Formation: As the crop is cut and deposited, it forms two main windrows, each containing two smaller sub-windrows. The sub-windrows are positioned side by side

within the main windrows, maximizing the surface area exposure to air and sunlight. This configuration facilitates faster and more uniform drying of the crop.

5. Drying and Conditioning: After formation, the crop undergoes a drying process facilitated by exposure to sunlight, air circulation, and natural evaporation. The increased surface area exposure resulting from the two-in-one double windrowing method enhances drying efficiency and promotes uniform drying throughout the crop. Farmers may monitor moisture levels and weather conditions to optimize drying efficiency.

6. Monitoring and Management: Throughout the windrowing process, farmers monitor the drying progress and field conditions to ensure optimal outcomes. This includes regular assessments of crop moisture levels, windrow integrity, and weather forecasts. Farmers may make adjustments to equipment settings or harvesting plans as needed to maximize drying efficiency and crop quality.

7. Subsequent Harvesting: Once the crop has reached the desired moisture level and is sufficiently dried, farmers proceed with subsequent harvesting operations using machinery such as combines or forage harvesters. The two-in-one double windrowing technique facilitates efficient harvesting by allowing equipment to gather and process the crop with minimal losses and maximum throughput.

As an example of windrowing operations, the following considers the process for harvesting canola as presented by Carmody (2009). Ensuring an even crop establishment is crucial for facilitating windrowing and harvest processes. It is preferred that stems have an average diameter of less than 10–15 mm for efficient operation of harvest machinery. Too few stems per square meter (<5/m2) can cause the windrow to sit on the ground, making it challenging to pick up, even with crop lifters. Additionally, excessively large plants can lead to uneven feeding into the harvester and difficulties in threshing. Planted row spacings of up to 50 cm typically do not pose a problem. Standard cereal width knife guards are recommended, and the knife must be in good condition to cut larger canola stems effectively.

The most common cutting height for canola ranges from 30 to 60 cm, depending on crop density and height. Harvesting crops too high can result in lower oil content and yield loss due to short branches falling to the ground. Some growers employ a chain or steel bar beneath the windrower exit to push tall stubble down before the windrow is laid, helping it to sit lower in the stubble and minimizing wind-induced movement.

A well-laid windrow is gathered evenly from both sides and interlocked with stems angled down at 45–60°, providing strength against the wind and facilitating even drying. Low-yielding crops (<0.7 t/ha) may be more susceptible to wind damage, in which case laying two windrows together using an end delivery windrower system can expedite the harvest process.

When harvesting, aligning the direction of harvest with the windrower reduces losses and minimizes the risk of crop stubble causing damage to the harvester's tyres. Key windrow settings include adjusting the reel to lightly push plants onto the windrower, maintaining a slightly faster reel speed than ground speed, and minimizing windrow blocking or bunching for smoother harvesting.

In windy conditions, a windrow roller can help anchor the windrow by pushing it into the stubble. Purpose-built windrow rollers can be attached behind the windrower and are effective in securing light windrows against wind.

The timing of harvest for windrowed canola is less critical than for standing crops. Windrowed canola can typically be harvested within 7–10 days after cutting, once the seed moisture drops to eight percent. Sampling the windrow within seven days and checking the moisture content aids in determining readiness for harvest.

All modern combine harvesters are suitable for harvesting canola, with open auger fronts or draper fronts being commonly used. However, belt pickup fronts are considered the best for harvesting windrowed crops like canola, as they gently pick up and feed the windrow into the harvester, reducing contamination and stubble material. Auger and draper fronts can be modified for windrowed canola harvest with adjustments to auger diameter, position, and reel type. Crop lifters can also be attached to open fronts as an alternative to belt pickup fronts, helping to pick up windrows efficiently.

Completing Windrowing Operations

Before departing from the operator's station, ensure several key steps are taken: Firstly, park the windrower on a flat surface to ensure stability. Next, shift the hydrostatic drive lever into neutral park position, ensuring the machine remains stationary. Disengage the platform drive to halt any movement of the cutting mechanism. Lower the platform to the ground to prevent any potential hazards. Turn off the engine and remove the key to prevent unauthorized use. Verify that the steering wheel is securely locked in place to prevent unintended movement. Finally, lock the cab door to secure the operator compartment. These precautions help ensure the safety and security of both the equipment and the operator when the windrower is not in use.

Completing windrow operations, cleaning equipment, and storing equipment are essential steps in maintaining agricultural machinery and ensuring its longevity. Here's a detailed explanation of each process:

1. **Completing Windrow Operations:**
    - Once the windrowing process is complete, the operator should shut off the equipment and safely disengage any moving parts.
    - Ensure that all windrows are properly formed and aligned for efficient drying and subsequent harvesting.
    - Conduct a final inspection of the field to check for any missed areas or irregularities in windrow formation.
    - Remove the windrower from the field and transport it back to the storage or maintenance area.
2. **Cleaning Equipment:**
    - Begin by removing any debris, plant material, or residue from the windrower using a high-pressure water hose or compressed air.
    - Pay special attention to areas prone to buildup, such as the cutterbar, conditioning rolls, augers, and belts.
    - Inspect the equipment for any signs of damage, wear, or malfunction that may require repair or replacement.
    - Lubricate moving parts according to the manufacturer's recommendations to prevent corrosion and ensure smooth operation.
    - Clean and inspect the engine air filter, radiator, and cooling system to maintain optimal engine performance.

- Check hydraulic fluid levels and top up as needed to prevent damage to hydraulic components.
- Remove any excess crop material from the platform and conditioning components to prevent clogging and reduce fire hazards.
- Clean the exterior of the equipment to remove dirt, dust, and other contaminants that may affect performance or appearance.

3. **Storing Equipment:**

- Park the windrower on a level surface in a designated storage area away from direct sunlight and exposure to the elements.
- Engage parking brakes and chock the wheels to prevent unintended movement.
- Lower the platform to the ground or engage the platform lift lockout lever to relieve tension on hydraulic components.
- Store the windrower under a protective cover or in a shed to shield it from dust, moisture, and UV rays.
- Disconnect the battery and store it in a cool, dry location to prevent discharge and corrosion.
- Secure all access panels, doors, and hatches to prevent unauthorized access and protect sensitive components.
- Consider performing routine maintenance tasks, such as changing fluids and filters, before storing the equipment for an extended period.
- Keep a record of maintenance activities, repairs, and any

issues encountered during the harvesting season for future reference.

CHAPTER FIVE

Disc Harrows

A disk harrow is a type of agricultural implement used for soil cultivation and preparation. It consists of a series of concave metal disks mounted on a common shaft or frame. These disks are typically made of hardened steel and are arranged in rows, with each row containing multiple disks spaced apart at regular intervals.

The primary function of a disk harrow is to break up and pulverize soil, incorporating crop residue, weeds, and other organic matter into the soil. This process helps to create a suitable seedbed for planting, improves soil aeration, and promotes water infiltration and drainage.

Disk harrows come in various sizes and configurations, ranging from small, single-row units suitable for garden plots to large, multi-row implements designed for large-scale agricultural operations. Some disk harrows feature adjustable gang angles, allowing operators to customize the aggressiveness of soil tillage according to specific field conditions and cropping practices.

Disk harrows are commonly used in conjunction with other tillage equipment, such as plows and cultivators, as part of a comprehensive soil management strategy. They are widely used in agriculture for preparing seedbeds, incorporating cover crops, managing crop residues, and controlling weeds, making them an essential tool for crop production worldwide.

Figure 39: John Deere tractor and disc harrow. John Deere tractor and disc harrow by Philip Halling, CC BY-SA 2.0, via Wikimedia Commons.

In comparing a disk harrow to cultivators and tillers:

Disk Harrow:

1. Function: A disk harrow is primarily used for breaking up and pulverizing soil, chopping crop residue, and preparing the seedbed for planting. It focuses on cutting and mixing the soil to create a suitable environment for seed germination and root growth.

2. Design: A disk harrow consists of multiple concave metal disks mounted on a common shaft or frame. These disks rotate as the implement is pulled through the field, cutting through soil and residue.

3. Depth: Disk harrows typically work at a shallower depth compared to cultivators and tillers, often penetrating only a few inches into the soil.

Cultivator:

1. Function: A cultivator is designed primarily for weed control and shallow soil cultivation. It is used to break up soil crust, control weeds between crop rows, and aerate the soil surface without significantly disturbing the soil structure.

2. Design: Cultivators feature multiple shanks or teeth arranged in rows, which penetrate the soil and disrupt weed growth. Some cultivators may also have attachments for hilling or ridging soil around crops.

3. Depth: Cultivators generally work at shallower depths than tillers, usually operating within the top few inches of the soil.

Tiller:

1. Function: A tiller, also known as a rotary tiller or rotavator, is used for deep soil cultivation, soil mixing, and seedbed preparation. It is more aggressive than a cultivator or disk harrow and is capable of breaking up compacted soil layers and incorporating organic matter to greater depths.

2. Design: Tillers consist of rotating tines or blades mounted on a horizontal shaft. As the tiller is driven forward, the rotating blades dig into the soil, breaking it apart and creating a finely tilled seedbed.

3. Depth: Tillers can work at deeper depths compared to disk harrows and cultivators, often reaching depths of several inches to a foot or more, depending on the model and settings.

In summary, while all three implements are used for soil cultivation and preparation, they differ in their primary functions, designs, and working depths. Disk harrows focus on cutting and mixing soil at shallow depths, cultivators are primarily used for weed control and shallow cultivation, and tillers are more aggressive and capable of deeper soil tillage and mixing.

Figure 40: Vogel & Noot compact disc harrow VN TerraDisc pro. © User:bdk CC BY-SA 3.0, via Wikimedia Commons.

A disk harrow works by using rotating concave metal disks to cut, chop, and pulverize soil, crop residue, and weeds. Here's how it typically operates:

1. Attachment and Power Source: A disk harrow is attached to a tractor or other suitable power source using a hitch mechanism. The tractor provides the necessary power to drive the harrow through the field.

2. Adjustment and Orientation: Before operation, the angle of the disk blades is often adjustable to control the aggressiveness of the tillage. The harrow is positioned at the desired depth and angle for the specific task.

3. Rotation of Disks: As the tractor moves forward, the disks mounted on the harrow's frame rotate rapidly. These disks can

be arranged in multiple gangs or rows, allowing for greater coverage and effectiveness.

4. Cutting and Mixing: The rotating disks cut into the soil, breaking up clods and slicing through crop residue and weeds. The concave shape of the disks helps to lift and chop the material, creating a fine tilth and mixing it thoroughly with the soil.

5. Soil Pulverization: As the disks penetrate the soil, they also pulverize it, breaking up compacted layers and creating a loose, friable seedbed ideal for planting. This action improves soil aeration, water infiltration, and nutrient distribution.

6. Coverage and Overlapping: To ensure thorough tillage, disk harrows are often designed to overlap slightly with each pass. This ensures that no areas of the field are missed and helps to achieve uniform soil preparation.

7. Depth Control: Depth control is typically achieved by adjusting the harrow's depth wheels or by altering the position of the hitch attachment on the tractor. This allows operators to set the desired working depth based on soil conditions and the specific requirements of the job.

Disk harrows consist of a series of large circular blades, known as disks, which are attached to a frame. These disks are typically made of steel and are arranged in rows along the width of the implement.

Here's how disk harrows work:

1. Attachment to Tractor: Disk harrows are usually attached to the back of a tractor via a three-point hitch. This hitch allows the harrow to be raised and lowered and adjusted for depth and angle.

2. Preparation: Before use, the harrow is adjusted according to the

specific requirements of the field. This may involve setting the depth of the disks to control how deeply they penetrate the soil.

3. Operation: As the tractor moves forward, the disk harrow is dragged across the field. The disks cut into the soil, breaking up clumps and chunks, and slicing through vegetation and crop residue. The rotation of the disks creates a mixing and pulverizing action, effectively breaking up soil aggregates and creating a smooth seedbed.

4. Adjustments: The angle of the disks can often be adjusted to control the aggressiveness of the tillage action. A more aggressive angle will result in deeper soil penetration and more thorough mixing, while a less aggressive angle will provide a lighter tillage action.

5. Coverage: Disk harrows are typically wide enough to cover a significant portion of the field with each pass. However, they may need to make multiple passes over the same area to achieve the desired level of soil tilth and smoothness.

6. Result: After the disk harrow has passed over the field, the soil is left smooth, level, and aerated, ready for planting or other agricultural operations. The tillage action of the disks helps to incorporate organic matter into the soil, improve water infiltration, and promote seed-to-soil contact for better germination.

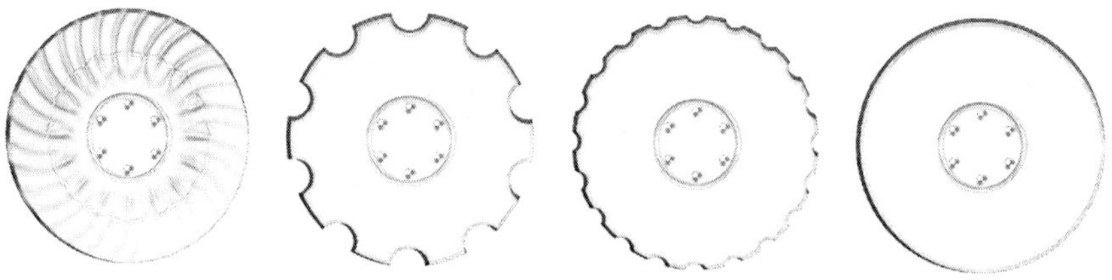

Figure 41: Spherical disks in harrow.

The primary mobile components of disc cultivators consist of spherical discs, typically with pointed edges along the perimeter, which may be either solid or notched. These discs are arranged into clusters, known as batteries, which are then mounted onto the frame in either single or multiple rows. When configured in a single row, they function as disc cultivators, whereas when arranged in multiple rows, they are termed disc harrows.

Discs are affixed to one, two, or multiple axles, which can be adjusted to various angles relative to the direction of motion. As the harrow advances, the discs rotate along the ground. Depending on the configuration of the discs, disc harrows are categorized into two main classes: Single action and Double action.

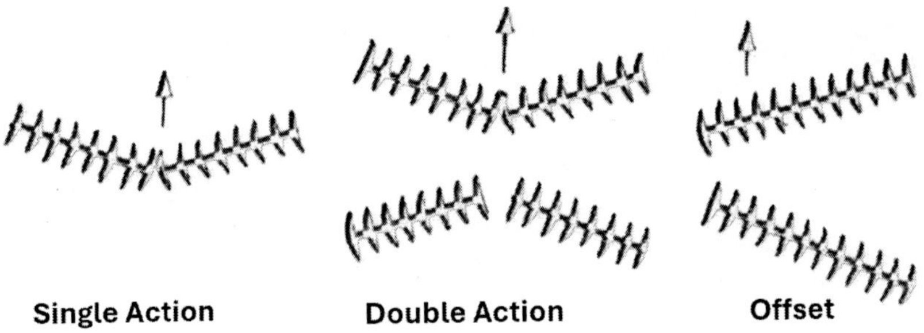

Figure 42: Types of disk harrow.

In a Single action disc harrow, two gangs are positioned end to end, dispersing soil in opposing directions. The discs are arranged so that one gang disperses soil to the right while the other gang disperses soil to the left.

Figure 43: Single action disk harrow. Михаило Јовановић, CC BY-SA 3.0 , via Wikimedia Commons.

Double action disc harrows feature two or more gangs, where one set follows behind the other. This arrangement ensures that the front and back gangs disperse soil in opposite directions, effectively covering the width of the field twice during each pass. Double action disc harrows are commonly available in two types: Tandem and Off-set.

Figure 44: Double action disk harrow. David Hawgood / Harrowing at Hampden Bottom, CC BY-SA 2.0, via Wikimedia Commons.

Tandem disc harrows comprise four gangs, each capable of being angled in the opposite direction.

Off-set disc harrows, on the other hand, consist of two gangs arranged in tandem, with the ability to be off-set to either side of the tractor's centreline. These harrows can cover widths ranging from 4 to 30 feet. Soil dispersal occurs in both directions due to the opposing orientation of discs on both gangs. Off-set disc harrows are particularly useful for orchards and gardens, as they can travel alongside the tractor and their off-set configuration allows for effective soil management. The design of Off-set disc harrows is based on the principle of counterbalancing side thrust, ensuring stability and efficient operation.

Additionally, the active segments are positioned at various approach angles ranging from 15 to 35 degrees. Increasing the approach angle results in both wider and deeper processing by each disc. For thorough stubble cultivation, a 35-degree approach angle is recommended, while for lightly weeded soil, a 30-degree angle suffices. Alternatively, a 15-20-degree angle is suitable for harrowing.

In recent times, advancements in disc harrow design have led to significant enhancements in functionality. Manufacturers now have the option to mount discs individually or complement them with additional mobile components such as rollers.

Machine tool constructions can be either trailed or mounted with a tractor.

A critical component of disc harrows is the bearing assembly, as the performance and longevity of the equipment hinge upon its quality. Ideally, manufacturers should utilize maintenance-free components to ensure hassle-free operation. Certain models of disc harrows are renowned for their maintenance-free design, eliminating the need for constant lubrication and saving farmers valuable time and resources.

Primary heavy-duty disc harrows, ranging from 265 to 1,000 pounds (120 to 454 kg) per disc, serve the primary purpose of breaking up uncultivated land, chopping up leftover material and residue, and integrating it into the topsoil. Secondary lighter disc harrows complement these tasks by thoroughly integrating residue from primary disc harrowing, removing clumps, and loosening compacted soil. Notched disc blades efficiently chop up stover from previous crops, such as cornstalks. By incorporating remaining residue into the topsoil, disc harrows facilitate the rapid decomposition of dead plant matter. Fertilizer application on residue atop the soil often results in a significant portion of applied nitrogen being absorbed by residual plant material, rendering it unavailable for seed germination. Disc harrows are typically employed before plowing to prepare the land for easier post-plowing management, reducing clogging and ensuring more thorough soil turnover during plowing.

Disc harrowing is the preferred method for incorporating agricultural lime (dolomitic or calcitic) and agricultural gypsum, achieving a well-balanced 50/50 mix with the soil when properly set, thus mitigating acid saturation in the topsoil and fostering robust root development. Given that lime doesn't naturally move within the soil, this presents a considerable challenge for sustainable zero-till farming, particularly considering the widespread use of chemical fertilizers by farmers globally.

Figure 45: Simba disc harrow. Evelyn Simak | Cultivated field north of Illington Road, CC BY-SA 2.0, via Wikimedia Commons.

In the event of wildfires, farmers commonly utilize disc harrows to swiftly create firebreaks between fields or around structures. This involves encircling a structure or field with the harrow, thereby tilling under flammable stubble, stover, or residue to deprive an advancing fire of fuel.

Secondary Uses: Once the heavy-duty discs wear down too small for further use in harrows, the hardened steel discs are repurposed to craft blades for hand tools used by wildland firefighters, farmers, and trail-building crews.

Primary tillage tasks are accomplished using heavy-duty discs equipped with large-diameter blades ranging from 26" to 40", along with increased disc spacings of 10", 14", and 18". These discs break virgin ground, integrate residue into the soil for ripper/subsoiler preparation, and alleviate soil compaction to enhance soil aeration and permeability at lower soil levels. Pre-planting operations often involve secondary disc harrows with narrower disc spacing ranging from 8" to 10" and disc sizes from 20" to 26". Alternatively, similar secondary tillage implements or rotary harrows are widely employed. Selection of secondary tillage equipment must consider soil type and moisture content at the time of use. Lighter secondary disc harrows primarily focus on breaking down soil clods into finer particles, promoting water penetration, enhanced soil aeration, and increased activity of soil organisms, resulting in a seedbed conducive to planting.

Disk harrow operations are vital for soil preparation in agriculture, yet they come with inherent hazards that can jeopardize the safety of operators and bystanders alike. Among the primary risks associated with disk harrow operations are entanglement and entrapment. The rotating disks of a disk harrow present a significant entanglement hazard, where loose clothing, long hair, or body parts can inadvertently become ensnared in the moving parts, potentially leading to severe injuries or fatalities. Similarly, individuals in close proximity to the harrow risk being pulled in and entrapped by the machinery.

Impact injuries are another concern during disk harrow operations. The high-speed rotation of the disks can propel debris, rocks, or other objects with considerable force, posing a risk of impact injuries to both operators and bystanders. These projectiles can cause lacerations, bruises, fractures, or even more severe injuries upon contact.

The heavy machinery involved in disk harrow operations also presents crushing and pinching hazards. Accidents may occur if operators or bystanders accidentally come into contact with moving parts or become caught between machinery components, leading to serious injuries or fatalities.

Tractors pulling disk harrows can overturn, particularly when operated on uneven terrain or slopes, resulting in significant risks for operators and passengers. Chemical exposure is another concern, as disk harrow operations often involve the use of fertilizers, pesticides, or herbicides, potentially leading to acute or chronic health effects through inhalation, skin contact, or ingestion.

Noise and vibration generated during disk harrow operations can cause hearing loss, musculoskeletal disorders, or other health problems for operators exposed to them over prolonged periods. Additionally, poor visibility conditions, such as dust, fog, or low light, increase the risk of collisions between the harrow and obstacles, other machinery, or personnel in the vicinity.

Prolonged operation of disk harrows can also lead to operator fatigue and ergonomic hazards due to prolonged sitting, repetitive motions, and awkward postures. To mitigate these hazards, operators should receive thorough training in safe operating practices, utilize personal protective equipment (PPE), conduct pre-operation equipment checks, and adhere to manufacturer guidelines and safety regulations. Implementing engineering controls such as safety guards and interlocks, along with proper equipment maintenance, can further minimize the risk of accidents and injuries during disk harrow operations.

Preparing for Disk Harrow Operations

Effective planning and preparation for disk harrow operations are crucial to ensure efficiency, safety, and successful soil preparation. Here's a comprehensive guide outlining the steps involved:

Firstly, it's essential to assess the field conditions thoroughly before commencing disk harrow operations. This assessment involves considering factors such as soil type, moisture content, presence of debris or obstacles, and terrain characteristics. By evaluating these conditions, you can determine the appropriate equipment and adjustments needed for optimal performance.

Next, select the right equipment for the job based on the field conditions identified during the assessment. Factors to consider include harrow size, disk diameter, and spacing between disks. It's imperative to ensure that the equipment is in good working condition and properly maintained to avoid any operational issues.

Familiarize yourself with safety guidelines and procedures for operating the disk harrow. Ensure that all operators receive proper training on safe operating practices, including the correct use of personal protective equipment (PPE) such as gloves, safety glasses, and hearing protection.

Develop a plan for the sequence of operations to be performed with the disk harrow. Determine the direction and pattern of harrowing, taking into account field shape, size, and accessibility. Plan for efficient turnaround points and transitions between passes to minimize downtime and optimize coverage.

Check the compatibility of the tractor intended for pulling the disk harrow with the equipment. Verify the tractor's horsepower, hitch capacity, and hydraulic system to ensure it can effectively operate the harrow. Make any necessary adjustments or upgrades to the tractor as needed for seamless integration.

Prior to starting operations, conduct a thorough inspection of the disk harrow and tractor. Check for signs of damage, wear, or malfunction, and address any issues promptly. Lubricate moving parts, adjust disk angles, and tighten bolts as needed. Ensure that all grease points are adequately lubricated, and hydraulic systems are functioning correctly.

Prepare the field by removing any large debris, rocks, or obstacles that could potentially damage the equipment or interfere with operation. Clear the field of any standing water or excessive mud to ensure smooth and efficient harrowing.

Based on the field conditions and operational plan, adjust the operating parameters of the disk harrow as needed. Set the gang angle, disk depth, and harrow speed to achieve the desired level of soil cultivation and residue incorporation.

Finally, if multiple operators or workers are involved in the harrowing operation, ensure clear communication and coordination among team members. Assign specific tasks and responsibilities, establish communication channels, and implement safety protocols to prevent accidents and ensure efficient operation.

Disk harrows need to be adjusted for several reasons to ensure optimal performance and efficiency during operation:

1. Field Conditions: Different fields may have varying soil types, moisture levels, and debris conditions. Adjustments are necessary to adapt the harrow to these conditions for effective soil tillage and residue management.

2. Depth Control: The working depth of the harrow affects its ability to penetrate the soil and incorporate residue. Adjustments in depth control ensure that the harrow operates at the desired depth, promoting proper seedbed preparation and soil aeration.

3. Penetration: Adjusting the angle of the gangs and other factors affecting penetration depth ensures that the harrow can effectively break up soil compaction and incorporate residue into the soil, promoting nutrient cycling and plant growth.

4. Leveling: Proper leveling of the harrow helps prevent uneven penetration and side draft, ensuring uniform tillage across the field and reducing strain on the tractor.

5. Scraper Adjustment: Scrapers on the harrow help remove debris and prevent buildup on the discs. Proper adjustment ensures that the scrapers function effectively, preventing clogging and maintaining optimal performance.

6. Attachment to Tractor: Proper attachment of the harrow to the tractor ensures stability and safety during operation. Adjustments in attachment points and alignment ensure that the harrow is securely connected to the tractor and operates smoothly.

Overall, adjustments in disk harrows are necessary to tailor the implement to specific field conditions, optimize performance, and achieve desired outcomes in soil preparation and residue management.

Adjustments for Disk Harrows:

a) Pre-Use Adjustments:

Before mounting the disc harrow, it is essential to ensure that all nuts and bolts are securely tightened. Additionally, a thorough assessment of the soil and debris conditions of the field is necessary to make initial adjustments. These adjustments include:

- Gang Angle Adjustment: The angle between two gangs can be adjusted within a range of 0° to 50°. Increasing the angle enhances penetration in dry soil, while decreasing it helps prevent clogging in wet soil.

- Disc Harrow Leveling: To prevent uneven penetration and side draft, utilize the top link and bottom adjustable link for leveling. When the tractor pulls to the right, slightly lower the rear gang, and raise it when pulling to the left.

- Scraper Adjustment: Loosen the bolts at the scraper's clamp to facilitate adjustment.

- Depth Control: Hydraulic control of the implement's working depth is achieved by raising or lowering the left control lever.

- Disc Harrow Penetration: Penetration depth is influenced by several factors including gang angle, harrow weight, disc diameter, disc sharpness (ensuring sharpness is crucial as blunt discs significantly increase draft), and hitch angle.

b) Attaching the Harrow to the Tractor:

When attaching the harrow to the tractor, follow these steps:

- Position the harrow on level ground to ensure stability.

- Reverse the tractor towards the harrow, avoiding dragging the harrow up onto the tractor.

- Begin by attaching the left arm of the tractor to the harrow.

- Connect the central top link/arm to the harrow, adjusting both screws equally until aligned with the central arm hole. If necessary, simultaneously adjust both screws to ensure proper alignment.

- Attach the lower right arm, adjusting the screw until the mounting pin aligns with the hole on the tractor arm. If needed, adjust both height and distance simultaneously. Once aligned, insert the pin and secure it with a lynch pin.

- After attachment, lift the harrow and ensure the control arm is parallel to the ground. When viewed from the rear or sides, all discs should uniformly touch the ground, indicating proper alignment and readiness for operation.

Operating Disk Harrows

Before initiating operations with the disk harrow, it's imperative not to position oneself between the harrow and the tractor, ensuring personal safety. Following this, properly fitting the three-point linkage as outlined earlier and securing it with a lynch pin is crucial for the stability and effectiveness of the setup. Should the scrapper come into contact with the discs during operation, it's advised to loosen the scrapper bolt and readjust its position. Additionally, it's essential to avoid turning the tractor sharply to the right or left while the harrow is engaged in the soil, as well as refraining from reversing the tractor under the same circumstances. To maintain optimal performance, discs should be replaced when their diameter decreases by 5" (125mm) from the original size.

Field Operation includes:

- To ensure effective independent soil breaking, lift the harrow while turning.
- Adjust internal and external check chains to maintain implement swing range within 50 mm (2") when raised.
- Maintaining correct tyre pressure is crucial to prevent wheel slippage during operation.
- When experiencing excessive rear wheel slippage, it is recommended to add wheel weights, water ballasting, or a combination of both.
- Correctly set hydraulic levers for draft and position control operations.

Side Draft: Equalizing the side thrust of the front and rear gangs ensures proper trailing behind the tractor. Adjusting the gang angle may be necessary to achieve this balance. ii) Severe Side Draft: Adjust the cutting depth of the rear disc gang using the tractor top link to mitigate severe side draft. Lower the rear gang when the tractor pulls to the right, and raise it when pulling to the left.

Driver Warnings: Before commencing harrowing, ensure all nuts and bolts on the harrow disc are securely fastened. Maintain vigilance regarding tree roots and stones, avoiding harrowing on stony soil. Additionally, ensure the tractor is in the appropriate gear and lift the disc harrow at every turn and before approaching the road for added safety.

Precautions During Transportation: During transportation, adjust the top link to the minimum length and set the hydraulic lever in the top raised position, locking levers securely. Maintain an appropriate speed to prevent jumping and exercise caution while overtaking on the road, displaying SMV symbols to indicate slow-moving vehicles.

Safety Symbols: Safety symbols, prominently marked on the disc harrow, serve as visual reminders for operational caution and safety adherence.

Operating disk harrows requires careful attention to safety, equipment setup, and execution of proper techniques. Below is a detailed guide on how to operate disk harrows effectively:

1. **Preparation and Safety Measures**:

    - Conduct a thorough inspection of the disk harrow and tractor before operation. Check for any signs of damage, wear, or malfunction, and address them promptly.

    - Ensure all safety guards and shields are in place and functioning correctly.

    - Wear appropriate personal protective equipment (PPE) such as gloves, safety glasses, and hearing protection.

- Familiarize yourself with the operator's manual for the specific disk harrow model being used and adhere to safety guidelines outlined within.

2. **Attach Disk Harrow to Tractor**:
   - Position the disk harrow behind the tractor on level ground.
   - Use the tractor's three-point hitch to attach the harrow securely to the tractor.
   - Ensure all attachment points are properly secured and locked in place.
   - Adjust the harrow's hitch height to ensure it is level with the ground when in operation.
3. **Adjust Operating Parameters**:
   - Set the gang angle of the disk harrow based on soil conditions. A greater angle may be required for more challenging soil types.
   - Adjust the working depth of the disks according to the depth required for soil cultivation and residue incorporation.
   - Set the harrow speed based on ground conditions and the desired level of tillage.
4. **Begin Harrowing**:
   - Start the tractor's engine and engage the power take-off (PTO) to activate the disk harrow.
   - Gradually lower the harrow into the ground to the desired working depth using the tractor's hydraulic system.

- Begin driving the tractor forward at a consistent speed, ensuring smooth and steady operation.
- Monitor the performance of the disk harrow, making adjustments to operating parameters as needed based on soil conditions and performance.

5. **Operational Techniques**:
   - Maintain a straight driving path to ensure even coverage and consistent tillage.
   - Avoid sudden stops or changes in direction to prevent uneven soil disturbance.
   - Monitor the depth of penetration and adjust as necessary to achieve the desired soil cultivation.
   - Keep an eye out for any obstacles or debris in the field that could damage the equipment and adjust course accordingly.
6. **Turnaround and Completion**:
   - When reaching the end of a pass, lift the disk harrow out of the ground using the tractor's hydraulic system.
   - Perform a smooth turnaround manoeuvre, ensuring ample space to safely reposition the harrow for the next pass.
   - Lower the harrow back into the ground at the desired working depth and continue with the next pass until the entire field is completed.
7. **Post-Operation Maintenance**:
   - Once harrowing is complete, thoroughly clean the disk har-

row to remove any accumulated debris or soil.

- Conduct a final inspection of the equipment for any signs of damage or wear and address any maintenance issues as needed.
- Store the disk harrow in a secure location, away from the elements, to prevent damage and ensure longevity.

By following these steps and adhering to safety guidelines, operators can effectively operate disk harrows to achieve optimal soil preparation and cultivation results. Regular maintenance and proper techniques are essential for safe and efficient operation of disk harrows.

Completing Disk Harrow Operations

The maintenance requirements of a disc harrow are influenced by factors such as the terrain it operates on. In stony terrain, for instance, the need for maintenance is heightened. If soil has infiltrated the grease nipple, it's advisable to replace the nipple promptly. Additionally, for newly acquired disc harrows, after the initial two hours of operation, it's essential to tighten all nuts and bolts to ensure optimal functionality. Following every fifty hours of use, it is recommended to grease all greasing points using a grease gun and thoroughly tighten all nuts and bolts. Moreover, after every fifty hours of use, it's crucial to open the bracket spool of the disc harrow, clean it with diesel oil, and replenish it with fresh grease.

Post-operation, it's imperative to wash the disc harrow to remove any accumulated debris or soil. Worn-out nuts and bolts should be replaced promptly to maintain the integrity of the harrow. If the disc harrow is anticipated to remain unused for an extended period, it's essential to thoroughly clean it and apply a layer of used oil to prevent rust formation, ensuring its longevity and performance in subsequent use.

Chapter Six

Seeders and Seed Drills

A seeder is a farming implement or machine used to sow seeds in the soil for crop production. It typically consists of a seed hopper, seed metering mechanism, seed tubes or rows, and a ground-engaging tool such as a disc opener or shoe.

The seed hopper holds the seeds, which are then released and metered out at a controlled rate by the seed metering mechanism. The seeds are conveyed through seed tubes or rows and placed into the soil at the desired depth and spacing by the ground-engaging tool as the seeder moves forward.

Factors influencing the amount of seed to be planted vary and often spark controversy, with a wide range of information, charts, and guides available due to differing environmental conditions across locations where forages are grown. Seeding rates also vary between forage and seed crops, necessitating familiarity with the seeding rate range in one's area. Notably, seeds have a high mortality rate, with predictions suggesting that only about one-third of sown seeds will result in seedlings, and only half of those will survive the first year. Despite this, the primary objective remains to seed enough for robust forage production, with seed expenses being relatively minor and high seeding rates potentially reducing weed growth.

A comprehensive seeding plan considers various factors, including the condition of the seedbed, soil moisture levels, expected germination percentage, species mixture if applicable, and the chosen seeding method.

Numerous methods exist for planting seeds, each with its own set of requirements for success and reliability in various conditions. Some methods involve the removal of existing plant growth through tillage or herbicide application, which not only prepares the soil but also addresses weed control. However, tillage may lead to soil moisture loss, erosion, and the introduction of new weed seeds to the surface. Conversely, no-till establishment, where existing plants are killed with herbicides and seeding is performed without tilling, offers an alternative approach but may require additional attention to fertility, pH, and water management.

Common seeding methods include broadcast seeding, grain drill with grass seed attachment, corrugated roller, band seeder (which adds fertilizer below and to the side of the seed), no-till, and aerial seeding. The objective across these methods is to achieve uniform seeding at a shallow depth (typically 1/4" to 3/8"), place fertilizer near the seed, and firm the soil around it. While most machinery accomplishes these tasks simultaneously, broadcast seeding may require separate rolling to ensure proper seed-to-soil contact, which can be achieved mechanically or by other means such as dragging heavy objects or allowing livestock to trample the seeded area.

Seeders come in various types and sizes, ranging from handheld seeders for small-scale gardening to tractor-mounted or trailed seeders for larger agricultural operations. They can be designed for sowing different types of seeds, including grains, vegetables, legumes, and cover crops, and may offer adjustable settings for seed spacing, depth, and placement to accommodate various crop requirements and field conditions.

A seed drill is a specialized farming implement designed for sowing seeds in rows at a consistent depth and spacing, representing a significant advancement in agricultural practices. This mechanized tool has revolutionized the process of planting seeds, leading to notable improvements in crop yields.

The key components of a seed drill include a seed hopper where seeds are stored before sowing, a seed metering mechanism to regulate seed flow, and seed tubes or rows that guide the seeds from the hopper to the ground. Additionally, a coulter or opener cuts a furrow in the soil to prepare the seedbed, while a seed delivery system ensures precise placement of seeds into the furrow through a chute or tube. Following seed deposition, a covering device such as a press wheel or drag chain covers the seeds with soil, facilitating optimal seed-to-soil contact for germination.

Seed drills are typically operated by being pulled behind a tractor and are available in various sizes and configurations to suit different crop types, field conditions, and planting methods. Their use enables farmers to sow seeds efficiently and accurately, leading to improved crop establishment, uniformity, and ultimately higher yields compared to traditional manual seeding methods.

Figure 46: Tractor with seed drill. Merilijoost, CC BY-SA 4.0, via Wikimedia Commons.

As such, seeders and seed drills are both agricultural implements used for sowing seeds, but they differ in their design, functionality, and the manner in which they plant seeds.

Seeders:

- Seeders are generally simpler in design compared to seed drills. They are primarily used for broadcasting seeds over a large area.
- Seeders typically scatter seeds evenly across the soil surface without creating furrows or rows. This method is suitable for planting crops like grass, cover crops, or certain grains.
- Seeders may have mechanisms such as a hopper and a metering device to regulate the rate at which seeds are released, ensuring even distribution.
- They are commonly used for overseeding lawns, pastures, or fields where crops are grown in a broadcast manner rather than in rows.

Seed Drills:

- Seed drills are more complex implements designed for precise seed placement in rows at a consistent depth and spacing.
- Seed drills create furrows in the soil using a coulter or opener, deposit seeds into these furrows at the desired depth, and then cover them with soil.
- They typically have components such as seed hoppers, seed metering mechanisms, seed tubes or rows, coulters, and covering devices to achieve accurate seed placement.
- Seed drills are commonly used for planting row crops such as grains, cereals, vegetables, and oilseeds where uniform spacing and depth are critical for optimal germination and crop growth.

A seed drill is an agricultural device designed to sow seeds for crops by accurately positioning them in the soil and covering them to a specified depth while being pulled by a tractor. This method ensures even distribution of seeds.

Figure 47: Seed Drilling. Seed Drilling on Elsham Hill by David Wright, CC BY-SA 2.0, via Wikimedia Commons.

By utilizing a seed drill, seeds are sown at the correct rate and depth, ensuring they are adequately covered by soil. This protects them from being consumed by birds and animals or drying out due to exposure to sunlight. Additionally, the seeds are arranged in rows, allowing plants to receive sufficient sunlight, nutrients, and water from the soil.

Before the advent of the seed drill, seeds were typically planted using hand broadcasting, a less precise and inefficient method resulting in uneven seed distribution and lower productivity. Implementing a seed drill can significantly increase the crop yield ratio (seeds harvested per seed planted) by up to eight times, while also saving time and labour.

Similar devices used for planting seeds are often referred to as planters, which have evolved from ancient Chinese practices into mechanisms that retrieve seeds from a bin and deposit them through a tube.

Many seed drills feature a hopper containing seeds positioned above a series of tubes, which can be adjusted to different distances from each other to optimize plant growth. The seeds are spaced out by fluted paddles, driven by a geared mechanism connected to one of the drill's land wheels. Adjusting gear ratios allows for changes in seeding rate. Modern drills commonly utilize air to transport seeds through plastic tubes from the seed hopper to the colters. This design allows seed drills to be much wider than the seed hopper, reaching widths of up to 12 meters in some cases. Seeds are mechanically metered into an air stream generated by an onboard fan powered hydraulically, then directed to a distribution head which divides the seeds into individual pipes leading to the colters.

Before operating a conventional seed drill, the hard ground must be plowed and harrowed to soften it sufficiently for proper seed placement and to create an optimal "seedbed" conducive to seed germination and root development. Plowing turns over the soil while harrowing smoothens it and breaks up any clumps. Alternatively, if the soil is not heavily compacted, it can be tilled with less invasive tools before seeding. The least disruption to soil structure and fauna occurs when using a drilling machine configured for "direct drilling," where seeds are sown into narrow rows opened by single teeth positioned ahead of each seed-dispensing tube, directly into or between the partially decomposed remains of the previous crop.

The seed drill must be adjusted according to the size of the seeds being used. Once set, the grains are placed in the hopper, from where they flow down to the drill, which spaces and plants them. This method remains in use today but has undergone updates and modifications over time, with notable advancements including wider machines capable of planting multiple rows simultaneously.

A seeder or seeding drill is typically attached to a tractor using a hitch mechanism. The hitch is a connection point located at the rear of the tractor, specifically designed to accommodate various implements such as seeders, plows, cultivators, and other agricultural machinery.

The attachment process involves aligning the seeder's hitch with the tractor's hitch point and securing them together using pins, bolts, or other fastening devices. Once properly connected, the tractor provides the power necessary to operate the seeding drill as it is dragged across the field.

Additionally, some seeding drills may require hydraulic or electrical connections to the tractor for functions such as raising and lowering the drill, controlling seed distribution rates, or engaging other operational features. These connections are typically made using hoses, cables, or electrical connectors that link the seeder's control systems to the tractor's hydraulic or electrical system.

An agricultural planter, also known simply as a planter, is a piece of farm equipment used for planting seeds in the soil at predetermined intervals and depths. It plays a crucial role in modern agriculture by efficiently and accurately placing seeds in the ground, facilitating crop establishment and growth.

Key components of an agricultural planter typically include seed hoppers or containers, metering devices to regulate seed flow, planting mechanisms such as seed tubes or discs, ground-engaging components like coulters or openers, and press wheels or packing wheels to ensure proper seed-to-soil contact.

Agricultural planters come in various types and configurations to suit different crops, soil types, planting methods, and farm sizes. Some common types of planters include:

1. Row Crop Planters: These planters are designed to plant seeds in rows with precise spacing and depth, suitable for row crops like corn, soybeans, cotton, and peanuts. They typically have multiple rows of planting units mounted on a toolbar or frame, allowing for efficient planting across large fields.

2. Drill Planters: Also known as grain drills, these planters are used for planting small grains such as wheat, barley, oats, and rye. They sow seeds in closely spaced rows or broadcast them across the field, depending on the seeding method and crop requirements.

3. Precision Planters: These advanced planters incorporate cutting-edge technology, including GPS guidance systems, variable rate seeding capabilities, and automatic row shut-off mechanisms. Precision planters enable farmers to optimize seed placement, spacing, and population to maximize yield potential while minimizing input costs.

4. No-Till Planters: No-till or minimum-till planters are specifically designed to plant seeds without prior soil tillage, preserving soil structure and minimizing erosion. They feature specialized openers or coulters that create seed furrows in untilled soil, allowing for direct seeding into residue-covered fields.

5. Vegetable Planters: These planters are tailored for planting vegetable crops such as tomatoes, peppers, cucumbers, and lettuce. They may include features for precision seeding, transplanting seedlings, or applying fertilizer and mulch during planting.

Figure 48: Row crop planters from 4 to 36 rows in size. IowaGoatWhisperer, CC BY-SA 4.0, via Wikimedia Commons.

While planters, seeders, and seed drills are all used for planting seeds in agricultural fields, they differ in their design, functionality, and the types of crops they are suited for. The differences are:

1. **Planters:**
    - Planters are designed to precisely place seeds in the soil at predetermined intervals and depths, typically in rows.
    - They are commonly used for planting row crops such as corn, soybeans, cotton, and sunflowers.
    - Planters often have multiple rows of planting units mounted on a toolbar or frame, allowing for efficient planting across large fields.
    - Many modern planters incorporate advanced technologies like GPS guidance systems, variable rate seeding capabilities, and automatic row shut-off mechanisms for enhanced precision and efficiency.

- Planters are suitable for both tilled and untilled soil conditions, depending on the specific model and farming practices.

2. **Seeders:**

    - Seeders are generally simpler in design compared to planters and are used for broadcasting seeds across a field rather than planting them in rows.
    - They are often used for sowing small grains such as wheat, barley, oats, and rye, as well as grasses and cover crops.
    - Seeders distribute seeds evenly over the soil surface, relying on subsequent cultivation or natural processes to cover the seeds with soil.
    - Seeder designs vary widely, ranging from hand-operated broadcast seeders to tractor-mounted broadcast seeders and air seeders capable of covering large areas quickly.
    - Seeders are commonly used in conjunction with tillage equipment or no-till drills to incorporate seeds into the soil.

3. **Seed Drills:**

    - Seed drills are specialized planting implements designed to sow seeds in rows at precise depths and spacing while minimizing soil disturbance.
    - They are particularly suited for planting small grains, legumes, and other row crops.
    - Seed drills typically consist of rows of planting units equipped with coulters or openers to create seed furrows, seed meters to regulate seed flow, and press wheels to ensure

proper seed-to-soil contact.

- Unlike planters, seed drills may not incorporate advanced technologies like GPS guidance or variable rate seeding.
- No-till seed drills are specifically designed to plant seeds without prior soil tillage, preserving soil structure and reducing erosion.

While all three types of equipment serve the purpose of planting seeds in agricultural fields, planters are specialized for precise row planting of crops like corn and soybeans, seeders are used for broadcasting seeds over a larger area, and seed drills are designed for row planting of small grains and legumes with minimal soil disturbance. Each type of equipment has its own advantages and is chosen based on factors such as crop type, field conditions, and farming practices.

Using seeders and seed drills in agricultural operations poses several hazards that workers need to be aware of and take precautions against. Some of the hazards associated with using seeders and seed drills include:

1. Mechanical Hazards:
    - Entanglement: Workers can get caught or entangled in moving parts, such as rotating shafts, belts, chains, or gears.
    - Crushing: There is a risk of being crushed between machinery parts or between the seeder and other objects.
    - Pinching: Workers may experience pinching injuries when adjusting or operating various components of the seeder.
    - Flying Debris: Loose parts or materials ejected from the seeder during operation can cause injury to nearby workers.
2. Chemical Hazards:

- Pesticide Exposure: Some seed drills may be used in conjunction with pesticides or herbicides, posing risks of exposure to harmful chemicals.

- Fertilizer Exposure: The handling and application of fertilizers can result in skin irritation or respiratory issues if proper protective measures are not taken.

3. Electrical Hazards:

    - Electric Shock: Electrical components in seeders and seed drills can pose a risk of electric shock if damaged or improperly maintained.

4. Ergonomic Hazards:

    - Awkward Postures: Prolonged use of seeders or seed drills may require workers to adopt awkward postures, leading to musculoskeletal disorders.

    - Repetitive Movements: Tasks such as loading seed or adjusting equipment settings may involve repetitive motions, increasing the risk of strain injuries.

5. Fall Hazards:

    - Falls from Height: Working on elevated platforms or climbing onto machinery to perform maintenance tasks can lead to falls and serious injuries.

6. Environmental Hazards:

    - Extreme Weather: Working outdoors exposes workers to weather-related hazards such as heatstroke, hypothermia, or sunburn.

- Terrain Hazards: Uneven or rough terrain can contribute to slips, trips, and falls, especially when operating seeders on slopes or inclines.

7. Noise Hazards:

    - Loud Machinery: The operation of seeders and seed drills can generate high levels of noise, potentially leading to hearing loss if adequate hearing protection is not worn.

It's essential for workers to receive proper training on the safe operation of seeders and seed drills, including hazard identification, risk assessment, and the use of personal protective equipment (PPE). Regular equipment inspections, maintenance, and adherence to safety protocols can help mitigate the risks associated with using seeders and seed drills in agricultural settings.

Preparing for Seeding Operations

Planning and preparing for seeding operations in agriculture is vital for ensuring the successful establishment and efficient growth of crops. To achieve this, it is essential to follow a comprehensive guide that encompasses various key steps:

- Assessing Environmental Conditions: It is crucial to evaluate weather forecasts and soil conditions to determine the optimal timing for seeding. Ideal soil moisture and temperature conditions are necessary to promote seed germination and early growth.

- Selecting Suitable Seeds and Varieties: Choosing crop varieties and seed types adapted to the climate, soil type, and intended end-use is important. Factors such as yield potential, disease resistance, and market demand should be considered when making these selections.

- Calculating Seeding Rates: Determining the appropriate seeding rates based on factors like seed size, germination rate, target plant population, and row spacing is essential. Seeking guidance from seed suppliers, agricultural extension services, or seeding rate calculators can aid in this process.

- Preparing Seedbeds: Ensuring that seedbeds are properly prepared to provide an optimal environment for seed germination and root development is critical. Using tillage equipment to cultivate the soil, remove weeds, and create a smooth, level surface helps in achieving this goal.

- Inspecting and Calibrating Equipment: Thoroughly inspecting seeding equipment, including seeders, drills, and planters, ensures they are in good working condition. Calibrating seeding equipment helps ensure accurate seed placement and consistent seeding rates.

- Planning Field Layout and Seed Placement: Determining the field layout, including row spacing, planting patterns, and seeding depth, based on crop requirements and equipment capabilities is necessary. Factors such as soil fertility, irrigation, and crop rotation plans should be taken into account.

- Preparing Seed Treatments (if applicable): Treating seeds with fungicides, insecticides, or other seed treatments as necessary to protect against pests, diseases, and environmental stressors is important. Following recommended application rates and safety precautions is crucial in this regard.

- Organizing and Storing Seeds: Properly organizing and storing seeds in a cool, dry location helps maintain seed quality and viability. Ensuring that seed containers are labelled accurately and securely sealed prevents contamination or moisture ingress.

- Arranging Labor and Resources: Allocating labour and resources effectively ensures timely completion of seeding operations. Coordinating equipment availability, manpower, and logistical support minimizes downtime and maximizes productivity.

- Developing a Seeding Plan and Schedule: Creating a detailed seeding plan outlining the sequence of operations, equipment requirements, and timelines for each field or crop is essential. Establishing clear communication channels and contingency plans helps address unexpected challenges or delays.

- Implementing Safety Precautions: Prioritizing safety during seeding operations by providing appropriate personal protective equipment (PPE), training workers on safe operating procedures, and adhering to regulatory requirements for equipment operation and chemical handling is crucial.

- Monitoring Progress and Adjustments: Continuously monitoring seeding operations, soil conditions, and crop emergence helps identify any issues or adjustments needed. Making timely modifications to seeding rates, equipment settings, or field management practices optimizes crop establishment and yield potential.

No-till or conventional drills feature various moving components, with each brand and model incorporating unique design elements that set them apart from others in the market. However, they typically share several common key features. In a no-till drill, a rolling coulter precedes the opener, slicing through sod, residue, and soil to create a slot, which is then widened by the double-disk opener. Conversely, conventional grain drills are utilized when the seedbed is already prepared, and the coulter is unnecessary for cutting through residue or the soil surface. As a result, conventional drills lack the coulter assembly at the front. Seed is contained within hopper boxes positioned atop the drill, with some models featuring multiple hopper boxes. Pasture-oriented drills typically include both large and small hopper boxes, with the former suitable for planting medium to large seed sizes (e.g., cowpea, pearl millet, tall fescue) and the latter for small-seeded species (e.g., clovers, brassicas, crabgrass). Additionally, certain drills may incorporate a "native grass box," specifically designed to accommodate native warm-season perennial grass species. These grass species often have fluffy appendages on their seeds, which can impede seed flow through the bottom of the hopper. Hence, native grass boxes are equipped with agitators to enhance seed flow within the box (Hancock et al., 2022).

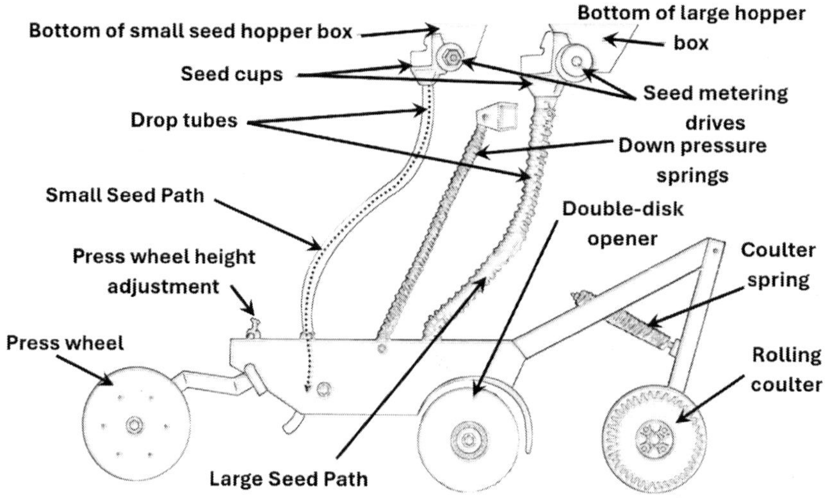

Figure 49: Typical components on one row unit of a no-till drill.

Unlike planters designed for row crops, seed drills operate differently by not separating and dropping seeds individually. Instead, seeds flow out of the bottom opening of a hopper box and are controlled by a seed metering unit, which adjusts the opening distance based on the seed size and desired seeding rate. As a driveshaft rotates, the seed is metered out and deposited into the seed cup.

As the drill moves across the field, a ground-driven wheel connected to a driveshaft rotates, allowing seed from a hopper box positioned above the row units to be metered into the seed cup. Subsequently, the seed descends through drop tubes and is released between or slightly behind the double-disks, effectively placing it within the furrow created by the opener in the soil. Following this, the press wheel trails behind, closing the furrow and compacting the soil around the seed. The weight of the drill compresses the down pressure springs, ensuring that the press wheels maintain contact with the soil surface.

Drills dispense a certain volume of seed rather than a precise weight or count of individual seeds. Therefore, when configuring and calibrating a drill, it's essential to consider factors such as the weight of seed per bushel (or other volume measure), the seeds' number per pound, the target seeding rate per acre in pounds, and the appropriate seeding depth. Frequent calibration, checking, and adjustment of the drill are necessary to accommodate changes in seed size or other planting conditions.

An example from the Southeastern U.S. illustrates the importance of accommodating different seed sizes, particularly evident with annual ryegrass seed variability (Hancock et al., 2022). Several varieties of annual ryegrass, particularly tetraploids containing four sets of chromosomes (4n), have larger seeds compared to the more common diploid varieties (2n). Despite having the same seeding rate in pounds per acre, adjustments to the seed metering unit opening may be necessary when planting diploid or tetraploid varieties to ensure the appropriate seeding rate is maintained.

In contrast to planters that precisely drop individual seeds into the furrow, drills disperse seed volumes more randomly, resulting in less precision. Consequently, counting seeds per distance of row may yield significant variation. Therefore, calibration of drills should focus on ensuring the recommended seed weight is distributed uniformly per unit area. Relying on the number of seeds per distance of row is not recommended for adjusting seeding rates or calibrating a drill.

Seeding rates for various crops can be obtained from several reliable sources:

1. Agricultural Extension Services: Local agricultural extension offices provide valuable resources and guidance for farmers, including recommended seeding rates based on regional conditions, soil types, and crop varieties.

2. Seed Suppliers: Seed companies often offer seeding rate recommendations for their specific seed varieties. They may provide this information on their websites, product catalogues, or through customer service representatives.

3. Government Publications: Government agencies involved in agriculture, such as the United States Department of Agriculture (USDA), publish guidelines and recommendations for seeding rates for different crops. These publications are often available online or through local agricultural offices.

4. Research Institutions: Universities and agricultural research institutions conduct studies and trials to determine optimal seeding rates for various crops. Research findings and recommendations are often published in academic journals, extension publications, or research reports.

5. Farming Organizations and Associations: Farmer associations and agricultural organizations may offer resources and publications that include seeding rate recommendations based on best practices and collective industry knowledge.

6. Online Resources: There are numerous online resources and databases where farmers can access seeding rate guidelines and calculators. These resources may include agricultural websites, forums, and apps designed specifically for crop management.

When using seeding rate recommendations, it's important to consider factors such as soil conditions, climate, crop variety, and farming practices specific to your region and farming operation. Additionally, periodically updating seeding rates based on field observations and local conditions can help optimize crop performance and yield.

Adjusting for the purity and viability of seed is essential due to variations in seed quality among different lots. Seeding rates provided in resources are typically based on high-quality seed and are expressed as pure live seed (PLS). Therefore, it's important to account for differences in seed purity and viability when determining the appropriate seeding rate. This can be achieved by calculating the PLS for each seed lot, which involves multiplying the percentage of viable seed (germination rate) by the seed purity percentage.

For instance, if a seed lot has 80% viable seed (total germination percentage) and 95% seed purity, the resulting PLS would be 76% (0.80 x 0.95 = 0.76). As a result, if the recommended seeding rate for a particular crop is 10 lb/acre (about 11.2 kg/hectare), one would need to adjust and apply 13 lb/acre (about 14.6 kg/hectare) using this seed lot (10/0.76 = 13).

Certain warm-season grass species naturally contain dormant seeds, which are viable but do not immediately germinate due to physiological factors. The percentage of dormant seed is included within the viable seed percentage when calculating PLS for seed lots containing dormant seeds. For instance, if a seed lot has 20% dormant seed, this percentage would be factored into the overall viable seed percentage.

In contrast, most annual species typically exhibit high germination rates and minimal levels of inert matter or weed seeds. Therefore, adjusting for PLS when establishing cover crops or annual forage crops is often deemed unnecessary or impractical, particularly if the germination rate exceeds 80% (Hancock et al., 2022).

Adjusting for seed coatings is a common practice among seed suppliers. Seed treatments, which often contain insecticides or fungicides, aim to protect seeds or seedlings from pests without altering seed weight or flow characteristics significantly. Conversely, seed coatings involve applying liquid binders and powdered fillers to coat the entire seed, altering its flow rate and sometimes carrying microbial inoculants, insecticides, or fungicides. Coatings containing Rhizobia inoculum for specific legume species have shown effectiveness, providing a convenient alternative to inoculating legumes before planting. However, the benefits of other seed coatings, especially those applied to grass seed, are subject to debate, with some studies showing marginal improvements in establishment success.

Certified-organic growers must be cautious, as most seed treatments and coatings are prohibited materials. They should opt for raw, untreated seed and inoculate legumes just before planting. Seed coatings can significantly influence the seed metering rate by a drill and add weight or volume to the seed, usually between 20-50%. Seeding rate recommendations and drill settings typically assume uncoated seed, necessitating adjustments to account for seed coatings. The weight added by the seed coating is listed on the seed tag as additional inert matter (Hancock et al., 2022).

Adjusting the seeding rate for pure live seed (PLS) allows for accounting for seed coatings. For example, if a seed lot of crimson clover has 85% viable seed (total germination percentage) and 49.85% purity, resulting in 42% PLS (0.85 x 0.4985 = 0.42), planting 16.3 kg of coated seed per hectare would be necessary if the recommended seeding rate for this crop is 6.7 kg/hectare (6.7/0.42 = 16.3). However, research findings on the necessity of adjusting for coated seed to PLS vary. While some studies suggest similar biomass production at only 50% of the PLS rate, others indicate increased biomass with PLS adjustment. Producers must weigh the potential risk of decreased biomass against the increased seed cost when considering PLS adjustment.

Drilling seed mixtures has become increasingly popular in establishing forage or cover crop stands, despite potential challenges due to differences in seedling vigor and seed sizes and shapes among species. Combining slow-establishing species with quick-establishing ones may lead to dominance by the latter. It's generally advisable to avoid simultaneous planting of annuals and perennials in forage, although mixing some perennials with annual species for cover crop plantings is feasible. Mixtures can serve various purposes, such as enhancing weed suppression and providing nitrogen while ensuring proper seeding rates is essential.

When planting grasses and legumes together, both large and small hopper boxes on the drill are utilized. Calibration of both hopper boxes independently is necessary to ensure accurate seeding rates. Complex mixtures with different seed sizes pose challenges in determining the seed metering unit's starting point for calibration. Segregation of seeds in the hopper box during drilling is another concern, especially with differing seed sizes. Periodically remixing the seeds in the hopper or filling it with an amount suitable for 1 to 1.5 hours of planting can mitigate this issue.

Ideal planting depth varies with species, complicating matters when planting mixtures. Compromising on planting depth for each species or using separate hopper boxes for large and small seeds can address this concern. Some drills allow routing drop tubes from the small hopper box in front of the press wheel to ensure proper seed placement. Adjusting planting depth based on seed characteristics is crucial for successful establishment of mixed seed plantings.

Prior to operating or calibrating a drill, ensure all preventive maintenance tasks are performed to ensure proper functionality. Preventive maintenance is often neglected, leading to stand failures, equipment malfunctions, and increased depreciation. Refer to the operator's manual for detailed equipment maintenance instructions and important safety information before calibrating the drill.

Key maintenance tasks include checking tyre pressure on all wheels, greasing all fittings, and lubricating the drive chains. Subsequently, inspect critical working parts of the drill, such as the seed metering units, no-till coulters, double-disk openers, and press wheels. Running a drill, especially no-till drills, over rough or rocky terrain can cause severe damage to components and hinder proper operation. Check for any damage, bending, or chipping on the coulters, openers, and press wheels. Replace damaged parts to ensure optimal performance. Double-disk openers should be replaced if their diameter is ½ inch less than the original diameter. An indicator of disk wear is if they can still hold an index card where they meet. It's vital that all parts run true with each other to facilitate proper furrow opening, seed dropping, and furrow closing for good seed-to-soil contact. This is especially crucial when using rented or commonly shared drills.

Additionally, maintaining the seed box and seed tubes is essential for proper seed distribution. Use a shop vacuum or compressed-air system to clean out old seed or debris. Insects like spiders and mud daubers may also build nests in the seed tubes, obstructing seed flow. After cleaning, test the drop tube with small pieces of paper to ensure it's clear. If there's any obstruction, remove the tube and use a flexible wire to clean out debris.

Operating a Seeder or Seed Drill

Complementing the hazards and risks associated with tractor operations, to mitigate the risks associated with using seeders and seed drills in agricultural operations, workers should implement the following precautions:

1. **Mechanical Hazards:**
    - **Entanglement:** Ensure all moving parts are properly guarded, and avoid wearing loose clothing or jewellery that could get caught.
    - **Crushing:** Maintain a safe distance from machinery in operation and use caution when moving around equipment.
    - **Pinching:** Use tools or equipment handles to adjust components rather than hands, and always follow safe operating procedures.
    - **Flying Debris:** Wear appropriate personal protective equipment (PPE), such as safety glasses and sturdy clothing, to shield against flying debris.
2. **Chemical Hazards:**
    - **Pesticide Exposure:** Follow pesticide handling instructions carefully, wear appropriate PPE, and ensure proper ventilation when working with pesticides.
    - **Fertilizer Exposure:** Wear gloves, goggles, and a respirator when handling fertilizers, and wash hands thoroughly after exposure.
3. **Electrical Hazards:**
    - **Electric Shock:** Regularly inspect electrical components for damage, and only operate machinery with intact wiring and grounded outlets.
4. **Ergonomic Hazards:**

- **Awkward Postures:** Adjust equipment settings to minimize the need for awkward postures, take frequent breaks, and perform stretching exercises.
- **Repetitive Movements:** Rotate tasks among workers to reduce prolonged exposure to repetitive motions, and use ergonomic tools designed to reduce strain.

5. **Fall Hazards:**
   - **Falls from Height:** Use fall protection equipment, such as harnesses and lanyards, when working at heights, and secure elevated platforms to prevent falls.
6. **Environmental Hazards:**
   - **Extreme Weather:** Schedule work during cooler parts of the day in hot weather, provide shade and hydration stations, and dress appropriately for weather conditions.
   - **Terrain Hazards:** Assess terrain conditions before operating machinery, use caution on slopes or inclines, and wear appropriate footwear for traction.
7. **Noise Hazards:**
   - **Loud Machinery:** Wear hearing protection, such as earplugs or earmuffs, when operating or working near noisy equipment, and limit exposure time to reduce the risk of hearing loss.

Various methods exist for calibrating a drill, typically outlined in the manufacturer's provided operator's manual. These instructions are typically tailored to the specific make and model of the drill. Some manufacturers offer calibration procedures that can be executed without moving the drill; instead, operators are instructed on how to activate or manually crank the seed metering unit while the drill remains stationary. While this method is generally satisfactory, operators should consider the terrain where the drill will operate. Uneven land or steep terrain can affect seed metering rates as the drill bounces or the seed shifts. Hence, it's advisable to perform a calibration or at least confirm seeding rates under actual operating conditions, especially on bumpy or hilly terrain.

For drills lacking calibration procedures in the operator's manual or for those preferring to calibrate under field conditions, a simplified calibration procedure is outlined below. This straightforward approach will guide users through calibrating virtually any drill in nearly any scenario.

Before beginning:

1. Ensure all preventive maintenance tasks are completed to ensure the drill's proper functioning.

2. Gather necessary supplies for calibration and adjustment:

    - Seed from the seed lot being planted
    - An accurate scale capable of weighing in grams, preferably to the nearest 0.1 g.
    - Calculator (optional, but recommended)
    - Wire flags or turf paint to mark a distance
    - Measuring wheel or 200-ft tape measure
    - Collection containers (e.g., plastic storage bags or small cups)
    - Cable ties and/or duct tape
    - Tool kit including adjustable pliers, screwdrivers, and a knife

Once the supplies are assembled, set the seed metering units to the specified opening size from the owner's manual or the chart on the hopper box lid. If the species or mixture being planted isn't listed, use the setting specified for a seed of similar size as a starting point. Ensure the seed cups are free of debris and set to meter seed. Fill the bottom of the hopper box for calibration so that each seed metering unit has at least 4 to 6 inches of seed covering the opening. Engage the drill and drive forward for 100-200 ft to ensure seed fills the metering units and is being metered into the seed cups.

Step 1: For each metering unit, detach the drop tubes from the metering unit below the hopper box or where it connects to the row unit. Ensure all units and drop tubes are clean and clear of obstruction. Securely attach plastic bags or cups to catch the seed below the seed cup or at the bottom of the drop tubes.

Step 2: Determine the width between row units either from the owner's manual or by measuring the distance between the centre of one row unit to the next. Use this distance to determine the calibration distance from the provided table.

Step 3: Use wire flags, turf paint, or another marker to measure and mark the calibration distance on typical terrain.

Step 4: With the drill's ground-driven drive mechanism engaged and traveling at the desired operating speed, start at the beginning of the calibration distance and travel to the end.

Step 5: At the end of the calibration distance and with the drill parked, collect the seed from the collection devices and record its weight using a scale that reads in grams.

Step 6: The grams of seed collected equals the pounds of seed sown per acre from that row unit.

Step 7: Adjust the seed metering unit opening as needed and repeat to ensure consistent seeding rates across all row units.

Adjusting planting depth is crucial for proper establishment of forage and cover crop species. It's essential to set up the drill according to the recommended planting depths as different species have varying requirements.

On a no-till drill, one can typically adjust planting depth by setting the cutting depth of the rolling coulter at the front of the drill. A general guideline is to set the coulter depth to be twice the desired planting depth. For instance, if the target planting depth is ½ inch, the coulter should slice 1 inch into the soil. This adjustment is usually done using a "depth control" knob or hydraulic system. Additionally, hydraulic stroke limiters or blocks can modify the cylinder's movement to prevent it from going deeper than necessary (Hancock et al., 2022).

However, merely adjusting the coulter depth is not sufficient. Other adjustments are also crucial, and these are similar for both no-till and conventional drill designs.

The depth of the double-disk opener is mainly determined by the downward pressure exerted by the weight of the drill and the down pressure springs. Typically, each row unit has one or two springs that apply downward pressure. Initially, these springs may be set to their lowest pressure setting by the manufacturer, which may be adequate at first. However, over time, these springs may lose tension and fail to provide sufficient downward pressure. To increase pressure, follow the manufacturer's instructions to shorten the spring's travel length, usually by adjusting the position of the "W" clip at the bottom of the spring to a higher hole in the rod.

Sowing and seeding techniques vary depending on the terrain and conditions. In situations where the land is flat and relatively obstacle-free, conventional, minimum-till, and no-till methods are suitable. Among these, conventional sowing is particularly effective as it ensures optimal soil-to-seed contact, a crucial factor for germination.

However, in areas with obstacles, such as rocks, aerial seeding and seed broadcasting may be preferable for efficient seed dispersal.

Proper seed placement and sowing depth are essential considerations regardless of the method used. Equipment should be accurately calibrated to meet the requirements of the specific pasture species, which can vary depending on factors like season and pasture variety. It's advisable to seek guidance from local seed suppliers or agronomists to ensure appropriate seed placement.

Undersowing a cereal crop with pasture can help offset pasture establishment costs. However, this approach requires careful management and favourable conditions to avoid compromising grain yield and pasture establishment.

Autumn is generally the best time for pasture sowing as it reduces the risk of failure due to heat and moisture stress that can occur with spring-sown pastures.

Conventional sowing typically results in superior pasture establishment, especially when weeds are effectively controlled and soil conditions are optimal. Cultivation and harrowing are essential for achieving a fine, firm seedbed, maximizing soil-to-seed contact. Post-sowing practices like using press wheels or harrows further enhance this contact.

Minimum and no-till sowing methods are effective in preserving soil structure, minimizing erosion, and reducing weed germination. However, achieving adequate soil-to-seed contact is more challenging than with conventional methods and requires regular equipment checks.

Broadcasting is suitable for areas where other sowing techniques are impractical, such as timbered regions or extensive grazing systems. While broadcasting can be effective, it often results in poor soil-to-seed contact and low seed germination rates. Additionally, seed-to-fertilizer placement may be inadequate.

Regardless of the chosen method, using fresh seed and following recommendations for seed inoculation and pest control are essential practices for successful pasture establishment.

Conventional sowing involves specific steps to ensure optimal pasture establishment, with an emphasis on controlling weeds and preparing a fine, firm seedbed to maximize soil-to-seed contact. Here's a detailed explanation of how to perform conventional sowing:

1. **Preparation of the Seedbed:**
    - Begin by preparing the field to create an ideal seedbed. This process typically involves plowing or tilling the soil to break up any compacted layers and create a loose, aerated seedbed.
    - Use implements such as a disc harrow or cultivator to further refine the seedbed, breaking up clods and leveling the surface. The goal is to create a smooth, finely textured seedbed free from large clumps or debris.
2. **Weed Control:**
    - Prior to sowing, it's essential to address any existing weed growth in the field. This can be done through mechanical methods such as plowing or tilling to uproot and bury weeds, or through the application of herbicides to suppress weed growth.
    - Effective weed control is crucial to minimize competition with the establishing pasture and ensure the success of the desired plant species.
3. **Seed Sowing:**
    - Once the seedbed is prepared and weeds are controlled, it's

time to sow the pasture seed. Use a seed drill or broadcast seeder to distribute the seed evenly across the field.

- Adjust the seeding rate according to the recommended specifications for the particular pasture species being planted.

4. **Maximizing Soil-to-Seed Contact:**

   - After sowing, it's important to ensure good soil-to-seed contact to promote germination and seedling establishment.
   - One method to achieve this is by using press wheels or harrows immediately after seeding. These implements can be attached to the seeder or pulled behind it to press the seed into the soil and improve contact.
   - Press wheels or harrows should be set at the appropriate depth to gently press the seed into the soil without burying it too deeply.

5. **Post-Sowing Management:**

   - Following sowing, monitor the field regularly for signs of emerging weeds or other issues that may affect pasture establishment.
   - Implement any necessary weed control measures to address weed growth that may emerge after sowing.
   - Additionally, provide appropriate irrigation if needed to ensure adequate moisture for germination and early seedling growth.

6. **Weed Management During Establishment:**

   - As the pasture seedlings begin to emerge, continue to monitor the field for weed growth.

- Implement timely weed control measures such as mowing, cultivation, or herbicide application to minimize competition with the developing pasture plants.
- Proper weed management during the establishment phase is critical to ensuring the success of the pasture stand and maximizing yield potential.

Implementing minimum and no-till sowing or seeding techniques requires careful attention to detail to ensure successful establishment of pasture while preserving soil structure and minimizing weed competition. Here's a step-by-step guide on how to perform minimum and no-till sowing:

1. **Field Preparation:**
   - Unlike conventional sowing, minimum and no-till methods do not involve extensive soil disturbance. Instead, the field is left largely undisturbed, with crop residues or cover crops often left on the soil surface.
   - Prior to sowing, assess the field conditions to ensure that the soil is not excessively compacted or covered with large debris that may impede seed placement.
2. **Weed Management:**
   - Weed control is essential in minimum and no-till systems to prevent weed competition with the emerging pasture plants.
   - Implement pre-plant weed control measures such as herbicide application or mechanical weed removal to address existing weed growth.

3. **Seed Sowing:**
    - Use a no-till drill or seed drill equipped with coulters or discs to sow the pasture seed directly into the untilled soil.
    - Adjust the seeding depth and spacing according to the specifications for the pasture species being planted.
4. **Soil-to-Seed Contact:**
    - Achieving good soil-to-seed contact is critical in minimum and no-till systems, as seed placement directly onto untilled soil can present challenges.
    - Regularly check the seeding equipment to ensure that seed and fertilizer placement are accurate and consistent.
    - Verify that press wheels or other mechanisms designed to assist with soil-to-seed contact are functioning properly. Adjust or replace components as needed to optimize performance.
5. **Monitoring and Adjustment:**
    - Monitor the field regularly after sowing to assess seedling emergence and growth.
    - Adjust seeding equipment settings or make necessary modifications to improve soil-to-seed contact if issues are observed.
    - Address any weed growth promptly to prevent competition with the emerging pasture plants.
6. **Post-Sowing Management:**

- As the pasture seedlings establish, continue to monitor the field for weed growth and other potential issues.
- Implement appropriate weed management strategies such as mowing, cultivation, or targeted herbicide application to control weed competition.
- Consider additional measures to promote pasture establishment and growth, such as irrigation or soil amendments if needed.

7. **Long-Term Maintenance:**
   - Once the pasture stand is established, continue to implement appropriate management practices to maintain soil health and productivity.
   - Rotate pasture species or implement cover cropping to further enhance soil structure and fertility over time.

Figure 50: Undertaking seeding operations, John Deere 7920, Kuhn seed drill combination. werktuigendagen, CC BY-SA 2.0, via Wikimedia Commons.

Operating a tractor with a seeder or seed drill attached necessitates meticulous attention to safety protocols to avert accidents and safeguard the well-being of the operator and those in proximity. Here's a detailed breakdown of how to safely operate a tractor with a seeder or seed drill attached:

Pre-Operation Inspection: Before commencing operation, conduct a comprehensive pre-operation inspection of both the tractor and the attached seeder or seed drill. This entails checking for any signs of damage, loose bolts, or worn components on the tractor, seeder, and hitch connections. Additionally, ensure that all safety guards and shields are securely in place.

Operator Training and Familiarization: Ensure that the tractor operator possesses adequate training and familiarity with the operation of both the tractor and the attached seeding equipment. Thoroughly review the operator's manual for both the tractor and the seeder to comprehend their functions, controls, and safety features.

Proper Hitching and Attachment: Utilize the appropriate hitch for connecting the seeder or seed drill to the tractor, ensuring a firm attachment. Double-check all hitch connections to verify that they are properly secured and locked in place before initiating tractor operation.

Weight Distribution and Stability: Pay close attention to the weight distribution of the tractor and the attached equipment to uphold stability and prevent tipping. Exercise caution against sharp turns or sudden movements that could destabilize the tractor, particularly when transporting a heavy load.

Safe Operation Practices: Adhere to wearing suitable personal protective equipment (PPE), including gloves, safety glasses, and hearing protection, throughout tractor and seeding equipment operation. Observe all safety decals and warning signs on the tractor and seeder, and comply with all safety guidelines outlined in the operator's manual. Maintain a safe distance from bystanders, structures, fences, and other obstacles while operating the tractor and seeding equipment. Be mindful of overhead power lines and other potential hazards in the work area, refraining from operating the equipment near them.

Operating Speed and Terrain: Operate the tractor at a safe and controlled speed appropriate for the terrain and prevailing conditions, especially when traversing rough or uneven terrain. Avoid operating the tractor on steep slopes or inclines beyond the recommended operating limits of the equipment.

Monitoring and Adjustments: Continuously monitor the performance of the seeder or seed drill during operation, making any necessary adjustments to ensure proper seeding depth, spacing, and coverage. Promptly halt tractor operation if any abnormalities or malfunctions are detected and address the issue before resuming operation.

Concluding Seeding Operations

After finishing seeding operations, it's essential to follow shutdown procedures to ensure safety and prevent damage to the tractor and seeding equipment. The process generally involves:

1. Lower the Seeder or Seed Drill to the Ground: Before turning off the tractor engine, lower the seeder or seed drill attachment to the ground. This action helps prevent any accidental movement of the equipment and minimizes the risk of damage.

2. Disengage All Moving Parts: Make sure to disengage all moving parts of the seeder or seed drill. This includes stopping any rotating components or mechanisms that were in operation during seeding. Disengaging these moving parts helps prevent injuries and damage to the equipment.

3. Securely Park the Tractor: Once the seeding equipment is safely lowered and all moving parts are stopped, securely park the tractor on level ground. Positioning the tractor on level ground ensures stability and reduces the risk of accidents or tip-overs.

4. Engage the Parking Brake: Activate the parking brake on the tractor to prevent it from moving unintentionally. Engaging the parking brake adds an extra layer of safety, especially when the tractor is parked on uneven terrain.

5. Turn Off the Ignition: Finally, turn off the ignition to shut down the tractor's engine completely. This step ensures that the tractor is no longer running and helps conserve fuel. Additionally, turning off the ignition reduces the risk of accidental starts and prevents unnecessary wear on the engine.

The seeder or seed drill should be stored in a designated area that provides protection from environmental elements and potential damage. Ideally, this storage location should be a secure and sheltered space such as a barn, equipment shed, or garage. Here are some reasons why the seeder or seed drill should be stored and secured properly:

1. Protection from Weather: Storing the seeder or seed drill indoors protects it from exposure to harsh weather conditions such as rain, snow, sunlight, and extreme temperatures. Exposure to these elements can lead to corrosion, rust, fading, and deterioration of mechanical components, reducing the equipment's lifespan and effectiveness.

2. Prevention of Theft or Vandalism: Securing the seeder or seed drill in a locked storage area helps prevent unauthorized access, theft, or vandalism. Agricultural equipment can be valuable assets, and storing them securely reduces the risk of theft or damage caused by intruders.

3. Safety Considerations: Storing the equipment in a designated area keeps it out of the way of pedestrian and vehicle traffic, minimizing the risk of accidents or injuries. Loose equipment left in open areas can pose tripping hazards or obstruct pathways, leading to safety concerns for workers and visitors.

4. Maintenance and Organization: Storing the seeder or seed drill in a controlled environment makes it easier to perform regular maintenance tasks and inspections. A designated storage area allows operators to keep track of maintenance schedules, spare parts, and tools, facilitating efficient upkeep and repair of the equipment.

5. Long-Term Preservation: Proper storage and securing of the seeder or seed drill help preserve its condition and functionality

over the long term. By minimizing exposure to environmental factors and potential hazards, the equipment remains in optimal working condition, ensuring reliable performance during future planting seasons.

Overall, storing the seeder or seed drill in a secure and sheltered location ensures its protection, prolongs its lifespan, enhances safety, facilitates maintenance, and preserves its value as an essential agricultural asset.

Chapter Seven

Sprayers

An agricultural sprayer, also known as a crop sprayer or pesticide sprayer, is a piece of equipment used to apply liquid substances such as fertilizers, herbicides, or pesticides to plants or crops. These substances are crucial for maintaining crop health and managing pests, diseases, and weeds throughout the crop growth cycle. Agricultural sprayers come in various sizes and types, ranging from handheld backpack sprayers to large trailed sprayers mounted on vehicles or tractors. They are designed to efficiently and accurately distribute liquid substances over crops to ensure optimal coverage and effectiveness.

Achieving uniform distribution of agricultural pesticides across crop foliage stands as the paramount objective. Underdosing risks insufficient coverage and control, while overdosing proves costly, leading to pesticide wastage and heightened potential for groundwater contamination.

Agricultural crop sprayers, also known as crop dusters or agricultural sprayers, are essential tools used in modern farming to apply various substances such as pesticides, herbicides, fertilizers, and fungicides to crops. These substances help control pests and diseases, promote crop growth, and improve yield. There are several types of agricultural crop sprayers, each with its own design, functionality, and application. Here are some common types:

1. **Boom Sprayers:**
    - **Description:** Boom sprayers are the most commonly used type of crop sprayers. They consist of a tank to hold the liquid chemical solution, a pump to pressurize the solution, and a boom equipped with spray nozzles that extend across the width of the sprayer.
    - **Functionality:** Boom sprayers are designed to deliver a uniform spray over a wide area, making them suitable for large-scale crop fields. They can be mounted on tractors or self-propelled vehicles, allowing for efficient application of chemicals.
2. **Airblast Sprayers:**
    - **Description:** Airblast sprayers utilize a powerful fan to create a high-velocity air stream that carries droplets of the chemical solution to the target crops. They often have adjustable nozzles and air deflectors to control the spray pattern.
    - **Functionality:** Airblast sprayers are ideal for orchards and vineyards where crops are densely planted and require precise application. They provide excellent coverage even in areas with dense foliage.
3. **Handheld Sprayers:**
    - **Description:** Handheld sprayers are portable devices that are manually operated by farmworkers. They typically consist of a tank, a pump mechanism, and a nozzle for spraying.
    - **Functionality:** Handheld sprayers are used for spot treatments, applying chemicals to specific areas or individual

plants. They are suitable for smaller farms, garden plots, or areas where precision spraying is required.

4. **Aircraft Sprayers:**
   - **Description:** Aircraft sprayers are mounted on airplanes or helicopters and are used for aerial application of chemicals over large agricultural areas. They feature high-capacity tanks and specialized spraying systems.
   - **Functionality:** Aircraft sprayers are effective for covering vast expanses of land quickly and efficiently. They are often used in areas with challenging terrain or when ground-based sprayers are impractical.
5. **Drip Irrigation Sprayers:**
   - **Description:** Drip irrigation sprayers deliver chemicals directly to the root zone of plants through the irrigation system. They consist of injectors that introduce the chemicals into the irrigation water.
   - **Functionality:** Drip irrigation sprayers are commonly used in greenhouse and drip irrigation systems. They provide precise dosing of chemicals and minimize wastage by delivering the solution directly to the plants' roots.
6. **Mist Blowers:**
   - **Description:** Mist blowers, also known as mist sprayers or foggers, produce a fine mist of chemical solution using a high-pressure pump or compressed air. They are often mounted on vehicles or carried by farmworkers.
   - **Functionality:** Mist blowers are used for applying chemi-

cals to crops with dense foliage, such as trees and bushes. They create a fog-like spray that can penetrate deep into the canopy, providing thorough coverage.

Each type of agricultural crop sprayer has its advantages and limitations, and the choice depends on factors such as the type of crop, field size, terrain, and application requirements. Farmers must select the appropriate sprayer for their specific needs to ensure effective and efficient chemical application while minimizing environmental impact.

Figure 51: Tractor and boom sprayer. , - CC BY-SA 4.0, via Wikimedia Commons.

An agricultural crop sprayer denotes the machinery employed for administering liquid substances onto plants or crops. These substances encompass fertilizers, herbicides, or pesticides, all pivotal for sustaining crop vitality throughout the growth cycle. The sprayers manifest in diverse sizes and configurations, ranging from handheld backpack sprayers to trailed variants.

An agricultural sprayer serves a multifaceted role, offering manifold benefits to the holistic management and well-being of crops:

Application of liquid fertilizers: By loading liquid fertilizers onto the sprayer, the equipment efficiently disseminates these nutrients to the crops. Its primary objective is to furnish crops with essential nutrients, leveraging its precision to target specific crops while mitigating loss through evaporation or environmental factors such as wind.

Water spraying: Water sprayers are instrumental in hydrating plants and crops. Operating at high pressure, they facilitate irrigation over localized areas. Additionally, preceding crop harvests, crops are subjected to water sprays to cleanse any residual chemicals.

Dispersion of herbicides or fungicides: These substances, when diluted with water, are sprayed onto crops to impede fungal growth or suppress unwanted vegetation.

Pest management: In pest control endeavours, crop sprayers wield exceptional efficacy. Pesticides, blended with water, are sprayed onto afflicted areas of crops to eradicate infestations.

For greenhouse pesticide application, two primary sprayer types are available: hydraulic and low-volume, each with various iterations tailored to specific crops or cultivation methods.

In a hydraulic sprayer, a pump propels the spray material—typically carried by water—to the target (plant foliage) at pressures ranging from 40 to 1000 psi. The spray, applied in a "wet" or "drip" manner, is broken into fine droplets by nozzles on the boom or handheld gun, directing it onto the foliage (Bartok, 2024).

Conversely, in a low-volume (LV) sprayer, spray material, suspended in a water or oil carrier, is introduced into a high-speed air stream generated by a fan, blower, or compressor. Often, a small pump injects a concentrated pesticide solution into the airstream, with air speeds reaching up to 200 mph. Achieving adequate coverage entails replacing the air within the foliage canopy with pesticide-laden air. Due to smaller droplet sizes, LV sprayers require less chemical for effective coverage.

Distinguishing between hydraulic and low-volume sprayers lies partly in droplet size. Hydraulic sprayers typically yield droplets ranging from 200 to 400 microns, while LV sprayers produce a mist (50-100 microns) or fog (0.05-50 microns). The minute droplets from mist or fog applicators offer more uniform coverage and increased likelihood of contact with pests or diseases. Unlike hydraulic sprayers, LV sprayers aim for a "glistening" appearance upon application, as individual droplets may be challenging to discern on the leaf.

However, smaller droplets evaporate swiftly in low humidity, potentially missing the target, and may bounce or skip on the leaf surface. This limitation can be mitigated by incorporating spreaders and stickers into the spraying process.

Figure 52: Self Propelled Sprayer. ShellieChatt, CC BY-SA 4.0, via Wikimedia Commons.

A hydraulic sprayer comprises several essential components: a tank, a pump with an agitator, a pressure gauge, regulating and relief valves, control valves, piping and nozzles, a power source, and a support frame.

Compressed Air Sprayer: Among the smallest types are hand-carried compressed air sprayers, featuring tanks with capacities ranging from 1 to 5 gallons (3.8 to 18.9 litres). These sprayers include an air pump positioned at the top, alongside a wand equipped with a nozzle for directing the spray. They are most effective for spot treating small areas. During operation, the tank requires frequent pumping to maintain pressure, and the chemical solution must be agitated by shaking the tank.

Backpack Sprayer: Equipped with a tank holding approximately four gallons of material, the backpack sprayer utilizes a hand-operated pump to pressurize the spray material as the operator moves along. The spray is directed to the target using a wand with a nozzle. Its utility is limited to treating small areas accessible from a walkway.

Figure 53: Farmer Geofrey Kurgat spraying his 1-acre of Korongo wheat against rust and army worm near Belbur, Nakuru. CIMMYT/ Peter Lowe, CC BY-SA-2.0, via Flickr.

Skid-Mounted Sprayer: Skid-mounted sprayers, featuring tank capacities of up to 200 gallons, can be mounted onto an ATV or electric cart. Alternatively, they may be mounted on wheels for manual pulling or attachment to a compact tractor. These sprayers are powered by a small electric or gas engine driving the pump. Some units may include a hose reel and gun, or a boom equipped with nozzles.

Irrigation Boom Sprayer: The irrigation boom sprayer has become invaluable for ensuring uniform watering, particularly in the context of increased production using plug and cell trays. By incorporating three-way turrets with nozzles for irrigation, misting, and pesticide application, this versatile equipment serves multiple purposes. Another approach involves adding a pesticide application boom to the same transport cart, necessitating an independent mixing tank, pump, filter, and valves.

Central Pesticide Application System: In gutter-connected ranges, a piping system can be installed to distribute pesticides throughout the greenhouse. Pesticide preparation and filtration are typically conducted in a designated mixing area. A single pump and piping capable of handling the developed pressure are essential. Each bay can be outfitted with one or more outlets to which a hose can be easily attached for pesticide application. However, one drawback is that the entire system must be drained and cleaned before switching to a different chemical.

Low-volume Sprayers include:

Backpack Mist Blower: Propelled by a small gas engine, a built-in fan generates an air stream with a velocity ranging from 100 to 200 mph. Concentrated spray is introduced into this air stream through a specialized nozzle and is carried to the foliage by the air. The spraying process is more intricate compared to hydraulic sprayers. It is essential to direct the nozzle into the plant canopy to ensure adequate penetration and coverage while maintaining a distance of at least six feet to prevent damage from the force of the air. The operator must ensure that the air within the canopy is entirely replaced by the air from the mist blower.

Electrostatic sprayer: Using compressed air, which acquires a negative electric charge as it passes through the nozzle, spray droplets are formed and directed towards the plants. This facilitates the creation of uniformly sized particles that disperse effectively due to their mutual repulsion. Charged particles are drawn to leaves, metal, and certain plastics; upon contact with a surface, they momentarily create an over-charge that repels other particles. This leads to more uniform coverage as other particles land elsewhere on the leaf.

The simplest electrostatic sprayer is portable and carried as a back-pack, equipped with a tank and spray gun. It requires an independent air supply to charge the tank. Alternatively, there are cart-mounted units with an integrated compressor powered by either a gas engine or an electric motor. Electrostatic sprayers perform optimally when the spraying distance is less than 15 feet.

Rotary Disk sprayer: This equipment employs a spinning disk to disrupt a stream of water into droplets ranging from 60 to 80 microns in diameter. Various sizes are available for use in greenhouses.

Thermal Fogger: Operating this device necessitates a carrier specially formulated to enhance droplet size uniformity and spray material distribution. The carrier also reduces molecular weight, enabling particles to remain airborne for up to six hours, which can be disadvantageous if immediate access to the greenhouse is required for plant care.

In operation, the pesticide is injected into a hot, high-velocity air stream, vaporizing it into fog particles. Covering the entire greenhouse can be achieved in as little as 15 minutes. For more uniform coverage and better foliage penetration, air circulation from a High Air Flow (HAF) system is recommended.

Temperature and humidity levels affect the behaviour of the spray droplets, and hearing protection is advisable due to the noise generated by the jet engine.

Mechanical Fogger: Also known as a cold fogger, this device utilizes a high-pressure pump (ranging from 1,000 to 3,000 psi) and atomizing nozzles to produce fog-sized particles. Spray material distribution can be achieved using a handheld gun or an external fan unit. The coverage area and distance depend on the fan's capacity, and multiple units or settings may be necessary for larger areas.

Similar to other foggers, penetration and coverage may not be as effective as with mist or hydraulic sprayers. Droplets measuring 30 microns drop out of the air relatively quickly, while those at 5 microns may evaporate or float in air currents for hours. Although small particles lack the mass or velocity to penetrate dense foliage effectively, studies have shown effective insect control in most cases.

A sprayer comprises various components such as a tank, pump, boom, and other elements for loading and cleaning the spray solution. However, the key components are the nozzles and a pressure gauge.

Nozzles:

- Nozzles play a crucial role in determining the spray quality (droplet size), the volume to be applied, the uniformity of the spray pattern, and the susceptibility to drift.

- Agricultural nozzles come in standard ISO sizes, typically with standard colours (e.g., blue 03), each with a specified output (e.g., 1.2 litres per minute at 3 bar). This facilitates easy interchangeability of nozzle sets or selection of low-drift versions without needing to adjust outputs, forward speeds, or pressures.

- Consult the product container label to ascertain the required water volume per hectare (usually between 100 and 200 litres/ha) and, if provided, the desired spray quality (droplet size), such as fine, medium, coarse, etc. This aids in selecting an appropriate nozzle size from the manufacturer's charts.

- While traditional spray charts offer guidance on forward speed, pressure, and nozzle size to achieve the desired application map, increasingly, user-friendly mobile applications enable real-time adjustment of settings in the field.

- Air induction nozzles represent the most common, effective, and straightforward type of drift reduction technology available. By incorporating air into the nozzle's fluid stream, these nozzles increase droplet size and significantly reduce drift, allowing safe spraying even under conditions that would typically lead to excessive drift with standard nozzles. Although air induction nozzles often produce larger droplets that may fall outside the parameters suggested on product labels, independent studies suggest that this typically has minimal impact on product performance and may be offset by the reduced drift. It is advisable to have both standard and air induction versions of the most commonly used nozzle sizes on hand to allow for switching based on prevailing conditions. Additionally, it's important to note that different brands of air induction nozzles may produce varying droplet spectrum sizes.

- Replace nozzles as a complete set if any signs of wear, such as increased flow rate (>10% higher than the rated capacity) or uneven spray patterns, are observed.

Pressure Gauge/Transducer and Forward Speed Sensor:

- An accurate and easily readable pressure gauge is indispensable for precise spraying. If a new nozzle yields incorrect output, it often indicates an inaccurate gauge or transducer.

- Accurate indication of forward speed is equally vital and should be regularly checked. Many basic GPS guidance monitors offer precise speed outputs.

Sprayer Control Systems:

- Many modern sprayers are equipped with flow transducers that enable calculation and control of the application rate per hectare. However, these systems rely on the accuracy of forward speed and flow transducers, which should be periodically verified.

- These control systems utilize system pressure to maintain a constant output even if forward speed fluctuates, which can affect spray drift and quality. Therefore, it remains essential to select the appropriate forward speed for the nozzle size and application rate.

Safety precautions are essential when using spray equipment with a high-pressure pump to prevent accidental skin penetration by spray particles.

Figure 54: A Househam Air Ride 3000 (AR3000) self-propelled sprayer at work (the sprayer has a 170 hp six cylinder engine, 3000 l tank capacity, 24 m booms. Dave Hitchborne / Crop Spraying, CC BY-SA 2.0, via Wikimedia Commons.

Spraying with a boom sprayer can pose various hazards, including:

1. Chemical Exposure: Direct contact with pesticides and other

chemicals used in the spraying process can lead to skin irritation, respiratory issues, eye irritation, and other health problems. Inhalation of pesticide vapours or accidental ingestion can also occur.

2. Drift: Spray drift refers to the unintentional movement of pesticide droplets away from the target area. Drift can result in the contamination of nearby crops, water bodies, residential areas, and wildlife habitats. It can also pose health risks to people and animals in the vicinity.

3. Equipment Malfunction: Malfunctioning or poorly maintained equipment can lead to accidents such as leaks, spills, or spray system failures. These incidents can result in chemical exposure, environmental contamination, and damage to the equipment itself.

4. Mechanical Hazards: Operating boom sprayers involves working with moving parts and machinery, which can pose risks of entanglement, crush injuries, or other mechanical accidents if safety precautions are not followed.

5. Environmental Impact: Improper application of pesticides can have detrimental effects on non-target organisms, including beneficial insects, pollinators, aquatic life, and soil microorganisms. Contamination of soil, water, and air can also harm ecosystems and biodiversity.

6. Noise Pollution: Boom sprayers often produce high levels of noise during operation, which can cause hearing damage or contribute to noise pollution in agricultural areas.

7. Ergonomic Issues: Prolonged use of boom sprayers may lead to ergonomic hazards such as repetitive strain injuries, mus-

culoskeletal disorders, and fatigue due to awkward postures or repetitive motions.

8. Chemical Spills and Accidents: Accidental spills or leaks of pesticides during handling, mixing, or application can result in environmental contamination, posing risks to human health, wildlife, and ecosystems.

9. Weather Conditions: Adverse weather conditions such as high winds, heavy rain, or extreme temperatures can impact the effectiveness of spraying operations and increase the risks of drift, equipment damage, or chemical exposure.

10. Regulatory Compliance: Failure to comply with local regulations, pesticide labels, and safety guidelines can result in legal consequences, fines, or penalties for operators and applicators. It's crucial to adhere to all relevant regulations and best practices to ensure safe and responsible spraying practices.

The components of an agricultural boom sprayer work together in a coordinated manner to ensure efficient and effective spraying of pesticides, herbicides, fertilizers, and other chemicals onto crops.

Firstly, the tank acts as the reservoir for holding the liquid solution to be sprayed. This solution is pressurized by the pump, which is a crucial component of the sprayer. The pump pressurizes the liquid, enabling it to be sprayed through the nozzles onto the crop canopy. This pressurization process ensures that the solution is atomized into fine droplets, which are essential for effective application.

The boom, extending horizontally from the sprayer, plays a vital role in facilitating uniform distribution of the spray over the crop. It holds the spray nozzles and ensures that the spray is evenly dispersed across the field. Booms come in various lengths and configurations to suit different application requirements and field sizes, ensuring optimal coverage of the crops.

Nozzles are integral to the spraying process as they atomize the liquid solution into droplets and direct them onto the crop. Different nozzle types can produce varying droplet sizes and spray patterns, which significantly impact the coverage and efficacy of the application. Proper selection and adjustment of the nozzles are essential for achieving the desired spray quality and coverage.

The pressure gauge is employed to measure the pressure within the sprayer system. Maintaining the correct pressure is crucial for achieving the desired spray coverage and droplet size, ensuring optimal effectiveness of the application.

Modern agricultural boom sprayers often feature control systems that enable precise adjustment of spray parameters such as pressure, flow rate, and nozzle selection. These systems may incorporate GPS technology for accurate application and automated control of spray sections, enhancing efficiency and accuracy.

In operation, the agricultural boom sprayer is filled with the desired liquid solution, and the pump is activated to pressurize the system. Spray parameters such as pressure and flow rate are adjusted based on the specific requirements of the application. The sprayer is then driven through the field, and the liquid solution is sprayed onto the crops via the boom and nozzles, ensuring even coverage and effective application. This integrated process ensures that the crops receive the necessary treatment for optimal growth and protection.

Spray drift requires careful management to mitigate its adverse impacts on the environment, nearby crops, non-target organisms, and human health. Several compelling reasons underscore the importance of managing spray drift effectively:

Environmental Protection: Spray drift can transport chemicals to unintended areas, contaminating soil, water bodies, and vegetation. This contamination may disrupt ecosystems, leading to ecological imbalances and potential harm to wildlife.

Crop Protection: Off-target drift poses a significant risk to neighbouring crops by exposing them to chemical residues. This exposure can result in reduced yields, financial losses for farmers, and potential disputes among landowners over damages.

Human Health Concerns: Inhalation or direct contact with chemical residues carried by spray drift can pose health risks to farmworkers, nearby residents, and consumers of affected crops. Managing drift is crucial to minimize these health hazards.

Regulatory Compliance: Compliance with regulations and guidelines governing pesticide application is essential for legal adherence and avoiding penalties or restrictions on chemical usage. Effective management of spray drift is a key aspect of regulatory compliance in agriculture.

Public Perception: Instances of spray drift incidents can tarnish the reputation of agricultural practices, leading to negative publicity and eroding public trust in food safety and environmental stewardship. Managing drift responsibly is vital for maintaining positive public perception.

Sustainability: Efficient management of spray drift contributes to sustainable agriculture by reducing chemical usage, minimizing environmental impacts, and promoting responsible stewardship of natural resources. It aligns with the principles of environmentally-friendly farming practices.

The effective management of spray drift is essential for promoting environmental sustainability, safeguarding human health, ensuring regulatory compliance, and fostering positive relationships within the agricultural community and broader society.

Some practical recommendations for minimizing off-target spray drift, which can result from various factors such as chemical volatility, weather conditions, equipment, and droplet size follow.

Start by thoroughly reviewing and adhering to the instructions provided on the product label, which often include specific guidelines regarding weather conditions, droplet size, equipment usage, and mandatory buffer zones. Following these instructions can help users effectively manage drift.

Before commencing spraying, assess the weather conditions to ensure suitability (wind speeds ranging from 3 to 15 km per hour blowing away from sensitive crops and areas, Delta T between 2 and 8, absence of an inversion layer). If the weather is unstable or unpredictable, refrain from spraying. Continuously monitor weather conditions during spraying, and cease operations if conditions become unfavourable.

Opt for chemical formulations less prone to off-target drift, such as amine formulations of 2,4-D instead of high-volatile ester formulations.

Survey the vicinity for susceptible plants, animals, and areas (such as streams and bee hives) near the target site, and implement strategies to shield them from drift. This may involve creating a buffer zone or leaving an unsprayed buffer adjacent to susceptible crops.

Engage in discussions with neighbouring properties, especially if spraying near sensitive crops or areas. This proactive approach allows them to take protective measures on their property if necessary and helps prevent potential complaints afterward.

Ensure that your equipment is properly set up and calibrated to ensure optimal performance.

Utilize a nozzle or sprayer setting that generates the largest droplet size possible (coarsest spray quality) without compromising the effectiveness of the chemical. Larger droplets are less likely to drift off-target.

Hydraulic nozzles are indispensable tools utilized in boom sprayers, particularly for treating expansive areas. These nozzles function by generating droplets when the spray liquid is pressurized and forced through a small opening. They are categorized based on the spray pattern they produce, with the main patterns being flat or tapered fan, even fan, hollow cone, and solid cone.

One of the crucial aspects of hydraulic nozzles is their ability to produce a wide range of droplet sizes, ranging from very small to large. Droplet size classification follows the ASAE Standard S-572, denoting categories from 'very fine' to 'extremely coarse,' with measurements taken at various percentages of the spray volume to ensure accurate classification.

Before selecting the appropriate nozzle size, certain parameters need to be considered, including the spray application rate, travel speed during spraying, and nozzle spacing. Calculating the required output per nozzle involves a formula that considers these parameters to ensure optimal performance.

In addition to selecting the right nozzle size, other factors such as nozzle filters and non-drip check valves play crucial roles in maintaining efficient operation and preventing blockages. Nozzle bodies come in different types suitable for dry or wet booms, with the latter typically involving tubing made of materials like PVC, polythene, or stainless steel.

Proper operation of boom sprayers with hydraulic nozzles is essential to achieve optimal results. Factors such as nozzle height, orientation, pressure, travel speed, and environmental conditions all influence the effectiveness of spraying. Manufacturers typically specify suitable operating heights based on the spray fan angle produced by the nozzles to ensure even coverage of droplets.

Adjusting nozzle height affects deposition uniformity across the boom, spray deposition on the target area, and the potential for drift from the sprayer. Similarly, altering nozzle pressure impacts both flow rate and droplet size, with higher pressure resulting in increased flow rate and decreased droplet size. However, it's crucial to adhere to the pressure range recommended by the manufacturer to ensure proper atomization and spray performance.

To determine the appropriate nozzle size, it's essential to have the following information:

- The spray application rate, measured in litres per hectare (L/ha)
- The travel speed during spraying, expressed in kilometres per hour (km/h)
- The spacing between nozzles, in meters (m)

Using the provided formula, you can calculate the required output per minute per nozzle:

Nozzle output (L/min/nozzle) = Spray application rate (L/ha) x Travel speed (km/h) x Nozzle spacing (m) ÷ 600

For example, if you aim to apply 40 litres per hectare at a travel speed of 20 km/h with a nozzle spacing of 0.5 meters, the calculation would be as follows:

Nozzle output = (40 x 20 x 0.5 ÷ 600) = 0.67 L/min/nozzle

Once you've determined the required output, you can select a suitable nozzle from the available range that delivers this flow rate within the recommended operating pressure range.

It's crucial to note the importance of nozzle filters in preventing blockages, particularly with low-volume nozzles. Typically, 100 mesh screens are used for nozzles with flow rates below 0.7 L/min, while 50 mesh screens are suitable for flow rates between 0.7 and 4 L/min.

Additionally, non-drip check valves are essential components to prevent herbicide from dripping from the nozzles after spraying ceases. These valves retain liquid in the spray line, facilitating prompt restart of spraying operations. Various brands offer diaphragm and ball-spring check valves, with all washers and diaphragms made from herbicide-resistant materials.

Nozzle bodies are available in two primary types, suitable for either dry or wet booms. Dry booms feature fixed nozzles connected by short lengths of flexible hose, while wet booms have nozzle bodies clamped around tubing, typically made of PVC, polythene, or stainless steel. Some nozzle bodies may accommodate multiple nozzles, allowing for easy rotation when frequent changes are necessary.

Planning and Preparing for Spraying

Sprayers are equipped with various filters, including suction, pressure, and possibly in-line filters, to prevent suspended materials from blocking the nozzles. Some air-induction nozzles are particularly prone to blockage and require correctly sized filters, especially at the nozzle level, to prevent clogging. Certain filter types, often found in nozzle bodies, offer a larger filter area and are less susceptible to blockages.

Cleaning the sprayer is crucial to prevent environmental contamination and avoid damage to the next crop. Internal sprayer washing systems should be utilized to clean residues from the sprayer, which can then be sprayed out in the field in a very dilute form. The extent of cleaning required beyond rinsing depends on the products used and the subsequent crop being treated. Special attention is needed when transitioning from herbicides in cereals to treating a broad-leaved crop such as beans, beet, or rape. This often requires multiple washings, including a complete fill and the use of a specific cleaning product.

Accurate calibration of the sprayer helps minimize excessive disposal, as the sprayer can be emptied in the field. Any remaining solution can be sprayed on the target field at a very low rate, provided it does not exceed the maximum allowable product rate per hectare.

To achieve optimal accuracy from a sprayer, it's essential to calibrate it before the spraying season begins and to recalibrate it regularly throughout the season. Applying too little pesticide can result in ineffective pest control, while excessive pesticide application not only wastes money but also risks damaging the crop and contaminating groundwater and the environment. Calibration aims to determine the actual rate of application in litres per hectare and make adjustments if the difference between the actual and intended rates exceeds 5%, as recommended by USEPA and USDA guidelines.

Before starting the calibration process, ensure that the sprayer has a good set of nozzles. Nozzles can wear out over time, leading to over-application or clogging. Clean and check the output of all nozzles for a specified time at a given spray pressure (typically 1 minute at 2.75 bar or 1 minute at 40 psi). Compare the output of each nozzle with the expected output from the nozzle catalogue at the same pressure. Replace any nozzles showing an output error exceeding 10% of the output of a new nozzle. Once this is done, the sprayer is ready for calibration. This includes (Deveau, 2017):

1. Fill the sprayer tank with water, ensuring it is at least half full.

2. Run the sprayer, check for leaks, and ensure all vital parts are functioning properly.

3. Measure the distance between the nozzles in centimetres/inches.

4. Determine an appropriate travel distance in the field based on the nozzle spacing.

5. Drive through the measured distance in the field at the normal spraying speed and record the travel time in seconds. Repeat and average the measurements.

6. With the sprayer parked, run it at the same pressure level and collect the output from each nozzle in a measuring jar for the

travel time determined in step 5.

7. Calculate the average nozzle output by adding the individual outputs and dividing by the number of nozzles tested. The resulting average nozzle output in millilitres corresponds to the application rate in litres per hectare.

8. Compare the actual application rate with the recommended or intended rate. If the actual rate deviates more than 5% from the recommended rate, adjust either the spray pressure, travel speed, or both. For example, to increase the flow rate, slow down or increase the spray pressure, and vice versa to reduce the application rate. Ensure adjustments are made within the proper and safe operating conditions of the sprayer, as increased pressure can lead to the production of small, drift-prone droplets.

9. Repeat steps 5-8 until the recommended application error of +5% or less is achieved.

For effective and safe spraying (Johnston, 2018):

Droplet Descent Time: Small drops take time to reach the target. A tiny droplet, measuring 100 microns in diameter (approximately 0.004 inches, or about the thickness of a human hair), descends 10 feet in 11 seconds. At 50 microns, it takes 40 seconds to cover the same distance due to the drag effect of air friction. This prolonged descent increases the risk of wind currents redirecting droplets to unintended targets.

These minute droplets experience significant deceleration upon leaving the nozzle. Attempting to increase pressure to propel these small droplets into the crop canopy is ineffective due to their rapid deceleration.

Droplet Size and Pressure: Droplet size varies with pressure changes. As sprayer pressure increases, droplet size decreases. Conversely, reducing pressure leads to larger droplets. Adjusting pressure during field operations alters droplet size, influencing how the spray material disperses in the air and potentially increasing spray drift. Therefore, it's advisable to reduce speed when spraying near neighbouring gardens or crops. Slowing down reduces nozzle pressure through the controller, resulting in larger droplets and minimizing drift risk.

Droplet Evaporation: Small droplets evaporate rapidly. Especially in warm temperatures typical of post-emergence applications, droplets under 150 microns evaporate within seconds under favourable conditions. Subsequently, wind can swiftly transport the chemical residue, leading to drift concerns. In contrast, larger droplets require two to three minutes to evaporate, ensuring they reach the target before significant water loss occurs.

Nozzle Variability: Nozzles yield a range of droplet sizes. Despite their similar appearance, sprayer nozzles produce varying droplet sizes. Each nozzle generates a spectrum of droplet sizes. For instance, when spraying a medium droplet ranging from 225 to 325 microns (approximately 0.009 to 0.013 inches), 5%-10% may consist of fine droplets under 150 microns, while another 5%-10% may comprise large droplets exceeding 450 microns. These fine droplets are most prone to drifting off target. It's prudent to replace aging nozzles exhibiting diminished performance characteristics.

Drift Dynamics: Drift increases with temperature and boom height. For example, at approximately 86 degrees Fahrenheit, a 10 miles per hour wind can induce a spray drift distance similar to a 15 miles per hour wind at 50 degrees Fahrenheit (24 kilometres per hour at 10 degrees Celsius). Additionally, a boom height positioned one to two feet above the crop canopy typically experiences minimal drift, whereas elevating it to three feet substantially increases drift distance.

Coverage and Efficacy: Prioritize coverage while considering droplet size. While larger droplets can reduce spray drift, it's essential to balance coverage and efficacy. Adequate coverage of the target area is crucial, and larger droplets may hinder this objective. For instance, a droplet size of 400 microns, combined with a spray volume of 15 gallons per acre, yields approximately 270 drops per square inch (6.7 litres per hectare yielding around 42 drops per square centimetre). This may suffice for systemic pesticides, which translocate within the plant. Conversely, reducing droplet size to 300 microns (at the same spray volume) increases coverage to 640 drops per square inch (99 drops per square centimetre), enhancing the performance of contact pesticides.

Label Instructions: Follow label instructions for optimal spraying techniques. For instance, when spraying fungicides on soybeans, achieving coverage on mid-plant surfaces is easier than on lower plant sections. Droplet size may not significantly affect coverage in this scenario, with fine, medium, and coarse droplets yielding comparable results. Some newer pesticides specify droplet size ranges on the product label, alongside the desired application rate. Adhering to label instructions provides valuable insights into effective spraying techniques and may address efficacy concerns later on.

Venturi Nozzles: Utilize Venturi nozzles to balance efficacy and drift. Newer Venturi-style nozzles offer advantages over older models by generating larger droplets at the same pressure. Equipped with a small hole drawing air into the liquid flow, these nozzles introduce air bubbles into the spray stream, resulting in larger droplets and reduced drift potential under certain conditions. Although the droplet size of newer Venturi nozzles is slightly smaller than older models, they maintain a balance between efficacy and drift based on application requirements.

Rate Controller Management: Manage rate controllers for consistent coverage. This ensures uniform application across the field. For instance, doubling the spraying speed from 6.2 miles per hour to 12.4 miles per hour (10 kilometres per hour to 20 kilometres per hour)

)necessitates doubling the flow rate to maintain the same spray volume, requiring a fourfold increase in pressure.

Operating a Sprayer

When using a self-propelled sprayer, before starting the engine, the initial step involves performing pre-start checks, which includes inspecting the essential fluids of your machinery such as fuel, hydraulic oil, hydrostatic oil, engine oil, and coolant levels. If you are using a self-propelled sprayer, it is likely that you will need to access the back of the machine to inspect some of these components.

Figure 55: Lite-Trac Crop Sprayer. Lite-Trac, CC BY-SA 3.0, via Wikimedia Commons.

Positioned at the rear of the sprayer, it's also essential to inspect the top of the tank to ensure its cleanliness and proper functioning of the tank rinse nozzles. Attention should then be directed towards the tyres. Utilize a pressure gauge to verify that all tyres are inflated to the correct pressure, adhering to the manufacturer's specifications. For trailed sprayers, it is crucial to verify the tractor tyre pressures as well (Deveau, 2017).

The aim is to maintain tyre pressure at the lowest recommended level for the load to be carried. This aids in boom height adjustment and stability, while also allowing tyres to function as shock absorbers to navigate bumps. In the case of trailed sprayers, use a spirit level to confirm that the drawbar is level. Adopt a systematic, clockwise approach to inspecting the machine to ensure thoroughness and avoid overlooking any components.

Moving on to the pumps, verify that they are adequately lubricated, ensure that toolboxes contain sufficient spare parts and necessary equipment, and confirm the presence of a spill kit equipped with absorbent granules and a spade in case of spillage. Lubricate all parts daily, and ensure that grease nipples are cleaned before and after use to prevent accumulation of dirt and blockages.

Inspect all hydraulic hoses, spray lines, and air lines for any signs of wear that may lead to operational issues. To detect leaks, it is advisable to operate the sprayer at a minimum pressure of 5 bar. Additionally, ensure that the spray tank is securely fastened, with all straps and bolts tightened.

Boom checks: Once the boom is deployed, assess its movement in the x- and y-axes. Given that each machine is unique, refer to the manufacturer's guidelines for proper boom setup. Verify that the tie rod closest to the rear of the machine exhibits slight looseness during movement, while the front rod remains tight. Next, assess vertical movement by gently pushing the boom down approximately 50cm and observing its return to the central position with minimal bouncing.

Maintaining the correct boom height is crucial during spraying, with the ideal height being 50cm above the crop. A practical method to ascertain this is by using a cable tie cut to the appropriate length as a visual aid from the sprayer cab. Ensure to measure from the tip of the nozzle to the crop rather than the spray line.

Given the importance of sprayer cleanliness, it is imperative to flush the system with clean water at the end of each day to eliminate any residue from the boom. In cases where the machine's boom lacks recirculation, periodic removal of the end caps and thorough flushing of the entire boom is recommended.

Nozzle checks: Verify that the nozzles are aligned both vertically and horizontally in accordance with NSTS guidelines. Conduct nozzle output checks at least twice annually by running the sprayer with clean water at a pressure of 3 bar. For an 03 nozzle operating for one minute at 3 bar pressure, the output should ideally be 1.2 litres/minute. Following output verification, cross-reference the figures with the rate controller to ensure correlation (Deveau, 2017).

Nozzle choice: The selection of a nozzle largely depends on the specific task at hand. A water volume of 100 litres/ha is recommended for spring fungicide application, providing sufficient coverage for effective disease control while optimizing sprayer efficiency.

Forward speed: The final aspect of minimizing spray drift involves controlling forward speed. Depending on nozzle size and water volume, aim to maintain a speed of 12kph (Deveau, 2017).

A tractor-operated spray pump utilizes the power of a tractor to drive its pump mechanism, constituting a cost-effective and straightforward spraying solution. Typically, it consists of a pump unit, a liquid solution tank, and hoses fitted with spray nozzles. The pump is connected to the tractor's power take-off (PTO), pressurizing the liquid before dispensing it through the nozzles for application. Compared to more complex mechanical tractor sprayers, these systems offer a simpler and more economical option for spraying tasks.

Mechanical tractor sprayers, alternatively known as boom sprayers or field sprayers, represent a more advanced and feature-rich option compared to tractor-operated spray pumps. They incorporate a diverse range of components and systems to enhance efficiency, accuracy, and convenience in spraying operations.

The tractor-mounted spray boom finds extensive use in various agricultural applications:

1. Crop Protection: Tractor boom sprayers play a pivotal role in crop protection strategies, where they are employed to apply pesticides, herbicides, and fungicides for controlling weeds, pests, and diseases. By ensuring precise and uniform distribution of these substances, they contribute to safeguarding crop health and promoting robust growth.

2. Liquid Fertilizer Application: These sprayers are also utilized for applying liquid fertilizers directly to crops. By delivering vital nutrients directly to the plants, they facilitate optimal growth and development. Adjustable flow rates and spray patterns enable precise and uniform distribution of fertilizers, maximizing nutrient absorption by the crops.

3. Field and Vegetation Management: Tractor boom sprayers are indispensable for managing field vegetation, enabling effective weed control and resource optimization. By selectively applying herbicides, farmers can suppress unwanted vegetation and promote the growth of desired plants, thus enhancing overall land management practices.

4. Orchards and Vineyards Spraying: In orchards and vineyards, tractor boom sprayers are commonly deployed for disease and pest control. These specialized sprayers are designed to navi-

gate through narrow rows and orchard/vineyard layouts, delivering fungicides, insecticides, or other treatments precisely to target areas. Adjustable boom heights facilitate reaching canopy levels of trees or vines, ensuring thorough and effective coverage.

Below are general recommendations or steps aimed at assisting pesticide applicators in achieving improved accuracy and efficiency in liquid pesticide application using boom-type ground sprayers (Ozkan, 2016). It's crucial for applicators to familiarize themselves with the specific recommendations for each pesticide and adhere to federal, state, and local government laws and regulations on pesticide application.

Review product label for specific recommendations and requirements:

- Equipment selection and setup: Identify any specifications for the sprayer, nozzle, spacing, pattern angle, travel speed, spray release height, or other factors listed on the label.

- Spray application rates (gpa): This information helps determine nozzle flow rates, influencing nozzle type, size, number, and operating pressure significantly.

- Spray classification/droplet sizes: Newer labels indicate minimum required droplet sizes as spray classification categories (e.g., very fine, fine, medium, coarse, very coarse, extremely coarse) to mitigate spray drift. Applicators select nozzles based on manufacturers' classification to match label requirements.

- Product agitation: Insufficient agitation may lead to issues with large or irregularly shaped tanks, particularly when applying wettable powder-type pesticides. Uniform mixing is essential, and some formulations have specific mixing requirements regarding constituents, order of addition, and agitation intensity.

- Adjuvants: Some labels specify the use of particular adjuvants to enhance product efficacy, influence droplet size or evaporation rate, reduce drift, and improve deposit and retention on the target.

- Application type: Certain pesticides may be highly volatile and require soil incorporation post-application. Following label recommendations is crucial to mitigate drift from volatile pesticides.

To identify acceptable application factors, it is essential to assess the operating capabilities of the sprayer through observation and consultation of operator manuals.

General inspection of the sprayer involves meticulously examining its components, including the tank, nozzles, hoses, pressure gauge, and pump, to ensure they are suitable in type and size and can operate effectively under various conditions. It's crucial to check for any leaks in the spraying system and ensure that pressure gauges return to zero when not in use. Additionally, verifying the efficiency of the tank agitation system and maintaining proper tyre pressure is essential to avoid variations in travel speed, which directly impacts the application rate. For sprayers equipped with automatic controllers, it's imperative to understand and properly operate the controller, valve settings, and programming.

Nozzle spacing plays a significant role in application effectiveness. Appropriate equipment setup and selection based on recommendations can optimize results for different types of spraying techniques such as banding, broadcast, or directed spraying. Nozzle spacing may need adjustments based on factors like crop type, spray application rate, release height, and pattern angle.

Consideration of nozzle flow rate and the number of active nozzles is crucial. Nozzles have limitations on flow rate and pattern, impacting the application rate per unit of boom width. Evaluating the sprayer's operating capabilities helps determine the maximum nozzle flow rate required, considering factors like pump capacity, pressure, and agitation flow requirements.

The spray pressure range directly affects the performance of the sprayer, influencing application rate and droplet size. Optimal pressure selection is vital to reduce drift and ensure adequate coverage. Similarly, sprayer speed range affects application rate, with higher speeds requiring higher nozzle flow rates. However, increased speed may also increase the risk of drift due to aerodynamic turbulence.

Maintaining appropriate boom height, nozzle spray pattern angle, and nozzle spacing is essential for uniform spray application and minimizing drift. Adequate size hoses and fittings are necessary to handle the pump output and total flow rate from the nozzles. Pump flow rate range and hydraulic agitation flow rates must also be considered to ensure sufficient liquid flow and agitation for effective pesticide application.

Consult the nozzle manufacturer's guide to:

- Explore the available nozzle types: There is a wide array of nozzles tailored for various applications. Selection of the appropriate nozzle type is influenced by factors such as application rate, spray pattern, and droplet size requirements. It's crucial to ensure that the chosen nozzle can deliver the required flow rate for achieving the desired application rate under the selected equipment settings. Ensure secure mounting of nozzles in their positions and maintain uniformity by using identical nozzles in all positions, except for fence row nozzles, which may be an exception.

- Determine the nozzle flow rate: Once the application rate, nozzle spacing, and travel speed are determined, the specific nozzle flow rate required can be calculated. Nozzles of different sizes within the same model may offer the necessary flow rate, as they typically have a wide range of flow rates, often by a factor of two or three. Selecting the appropriate nozzle size and operating pressure to produce droplets that minimize drift is typically feasible.

- Assess the spray pressure range: A specific nozzle size may yield different flow rates depending on the operating pressure. Opting for a nozzle size that delivers the specified flow rate at a lower operating pressure often helps reduce drift given the equipment setup and operating conditions.

- Examine the nominal spray angle: Nozzles are available with various spray pattern angles. Wider spray pattern angles enable lower spray release heights, ensuring adequate spray pattern overlap and improved coverage for a given nozzle spacing. Increased nozzle spacing necessitates a broader spray pattern angle to achieve full overlap. Product label specifications regarding spray pattern overlap also influence the required nominal spray angle of the nozzle. Limited nozzle spacing may dictate the necessary spray pattern angle for a given range of acceptable boom heights.

- Verify the spray classification/droplet sizes: Confirm that the selected nozzle, operating at the specified pressure, produces the desired range of droplets for enhanced efficacy and reduced drift. Label requirements may mandate nozzles that generate a specific droplet size spectrum, such as fine, medium, coarse, or very coarse. Carefully review the label to identify the optimal droplet size and suitable nozzle size for the intended spray application. Minimize drift by ensuring that the nozzle generates a minimal volume of smaller droplets (typically less than 150 to 200 microns) at the specified operating pressure.

Determine the sprayer setup to achieve an acceptable application rate:

Choose the nozzle spacing: Once a suitable general nozzle type has been identified to generate the required flow rate with a satisfactory droplet size, the nozzle spacing can be chosen. This selection depends on factors such as the spray pattern angle and desired boom height, ensuring proper overlap and sufficient coverage. In cases where drop nozzles are utilized, the spacing is determined by the crop row spacing.

Select the spray rate: The application rate, typically specified as a minimum requirement on the label, is chosen based on label instructions. However, the actual application rate may exceed this minimum. After selecting the application rate, the nozzle flow rate, nozzle spacing, and sprayer travel speed dictate the actual application rate.

Determine the sprayer speed: Identify the optimal sprayer travel speed considering label specifications, sprayer capabilities, terrain conditions, aerodynamic effects, and boom stability requirements to minimize drift.

Calculate the required nozzle flow rate: Calculate the required nozzle flow rate based on the chosen nozzle spacing, spray application rate, and sprayer speed using the formula: NFR (gpm) = AR (gpa) * TS (mph) * NS (in) / 5940. Ensure accurate measurement of travel speed in the field and avoid relying solely on tractor or sprayer instruments for speed determination.

Select the specific nozzle type and size: Choose the appropriate nozzle tip size considering spray classification and nozzle flow rate requirements. Ensure that the selected tip has the capacity to deliver the required flow rate and produce the desired pattern and droplet spectrum as per label instructions.

Determine nozzle tip operating pressure: Once the nozzle tip is selected and confirmed to match the hydraulic pump's flow and pressure capacity, determine the precise operating pressure required to achieve the desired flow rate based on the nozzle tip operating pressure/flow rate specifications provided.

Confirm nozzle tip spray classification: Verify that the selected nozzle tip indeed produces the desired spectrum of droplets at the chosen operating pressure, using manufacturer specifications or reputable droplet measurement instruments.

Ensure alignment with label classification: Compare the actual droplet spectrum produced by the nozzle with the labelling or other requirements for droplet size. If the nozzle generates an excessive volume of fine droplets, repeat the selection process until an acceptable droplet spectrum is attained.

Set boom height based on nozzle angle and spray overlap: Confirm the appropriate boom height to achieve the desired overlap with the selected nozzle and tip's spray pattern angle. Opt for the minimum applicable boom height to minimize spray drift.

Calibration is indispensable to ensure that a sprayer applies chemicals at the recommended rate accurately. For safety reasons, calibration should be conducted using water as the spray solution. Various calibration procedures are available from nozzle or sprayer equipment manufacturers and local Extension offices.

Regular calibration is essential as the flow rate of nozzles, particularly those made from brass, tends to increase as they wear out. Research from the University of Nebraska demonstrates a direct correlation between the number of calibrations performed by the applicator and spraying accuracy. Most modern sprayers are equipped with automatic controllers, and their accuracy must be verified as part of the calibration process.

Confirm flow rate: Measure the nozzle output flow rate over a specified time period. All nozzles should maintain a flow rate within +/- 10% of the original output at the given pressure. Any nozzle falling outside this range should be replaced.

Verify pressure: Maintaining the correct pressure at the nozzle is crucial for achieving the desired flow rate. Pressure readings from spray rate controllers or gauges near pressure relief valves often reflect pressure closer to the pump, not at the nozzle. Hence, it is advisable to use a second pressure gauge to check the pressure at the nozzle.

Validate travel speed: Ground speed significantly influences the application rate. Doubling or halving the travel speed results in corresponding changes to the application rate. Measure the true ground speed by driving a set distance in the field at the normal spraying speed and recording the travel time. Calculate the travel speed in miles per hour using the distance and time measurements.

Ensure uniformity: Maintain consistent deposition of spray material across the boom to ensure uniform coverage on the target. Non-uniform coverage can occur due to various factors such as misaligned or clogged nozzles, differing fan angles, or uneven nozzle height. Using a portable patternator can help assess spray uniformity.

Mixing chemicals:

- Assessing Product Compatibility: When mixing two or more pesticides for spraying, it's crucial to consider the potential risks of undesirable outcomes. Even if the products are compatible, improper mixing order can lead to issues such as increased density or viscosity of the tank mixture, reduced efficacy due to product antagonism, or separation in the tank. Pesticide labels usually provide detailed guidance on compatibility and the sequence of mixing with other products.

- Determining Chemical Amount: To ensure uniform coverage over the sprayed area, pesticides must be mixed in an adequate liquid carrier. Applicators should accurately calculate the amount of chemical needed to meet label requirements. The dilution rate must align with the instructions on the label.

- Estimating Tank Capacity: For precision, it's essential to determine the exact volume of the carrier solution in the tank before adding chemicals. Verify the accuracy of liquid volume markings on the tank exterior. Inaccurate markings can result in underapplication or overapplication of the active ingredient. Tank capacity measurement can be done using a flow meter or by weighing the sprayer on a scale when empty and full, then converting the weight difference to gallons of water.

- Water Cleanliness and pH: Use clean water free from foreign materials to avoid clogging screens and damaging nozzles and pumps. Water from natural sources like ditches, ponds, or lakes should be filtered before filling the tank to prevent contamination. Additionally, consider the pH of the water, as it can affect the efficacy and stability of certain chemicals. Refer to the chemical label for recommendations on adjusting water pH to the desired level.

Spray additives encompass a diverse range of products designed to fulfill various functions, including mitigating droplet evaporation and drift. Many labels for product formulations specify the use of particular adjuvants and provide detailed instructions for their mixing and agitation. These instructions often consider solution properties that influence droplet evaporation and drift, such as specific gravity and fluid characteristics, which may differ from those of water used for calibration. Adjustments may be necessary to accommodate these differences.

The effectiveness of drift reduction with adjuvants can vary. While some may lead to an increase in relative droplet size span, others might simultaneously enlarge larger droplets while diminishing smaller ones. Certain formulations and adjuvants may generate long-chain polymers that could be susceptible to shearing by pumps, thereby reducing their efficacy in decreasing the volume of spray contained in smaller, drift-prone droplets. It's important to note that drift-retardant chemicals should be viewed as a secondary measure against drift, to be employed after considering other drift mitigation strategies such as utilizing low-drift nozzles, opting for nozzles with larger orifices, or reducing the spray pressure.

Determining the optimal timing for pesticide application involves careful consideration of various factors to ensure effective and efficient treatment while minimizing potential risks. First and foremost, it's essential to assess whether the level of pest infestation justifies pesticide application economically. This assessment involves scouting the target area and evaluating pest threshold levels before deciding to spray. If conditions do not warrant immediate application, it may be prudent to wait for a time when the pesticide will offer the highest level of control, or alternatively, to focus spraying efforts on specific infested patches, a strategy that may be more feasible for weeds compared to insects and diseases.

Before initiating spraying activities, a thorough site assessment is crucial. This assessment involves gathering detailed information about the surrounding environment through prior planning, on-site visits, aerial maps, and discussions with neighboring stakeholders. Understanding the surrounding terrain and potential sensitive areas informs decisions regarding spray timing, suitable weather conditions, necessary equipment, and drift mitigation techniques.

To minimize spray drift, which accounts for a significant portion of non-compliance cases investigated by regulatory authorities, various proven strategies can be adopted. These include selecting appropriate nozzles that produce coarser droplets for targets that do not require small droplets, maintaining higher application rates, utilizing low-drift nozzles, and employing air-assisted or electrostatic sprayers designed specifically for drift reduction. Additionally, factors such as spray pressure, weather conditions, and atmospheric stability must be carefully considered to further mitigate drift risk.

Recording accurate and detailed information during spraying operations is essential for regulatory compliance and potential litigation purposes. This includes documenting wind speed and direction, as well as nozzle types, sizes, spray pressure, and environmental conditions. Following safety guidelines outlined in sprayer operator manuals and chemical labels is paramount to ensure the safe handling and application of pesticides, which may include wearing appropriate protective clothing and equipment.

Proper management of pesticide waste and empty containers is critical to prevent environmental contamination. Practices such as reducing pesticide consumption, rinsing containers promptly, and participating in container collection programs can significantly reduce waste. Before commencing spraying, it's imperative to conduct a final check to ensure all sprayer components are functioning correctly and securely fastened.

During spraying, periodic checks of nozzle output patterns are necessary to detect any issues such as clogging or foreign materials. Regular maintenance and cleaning of the sprayer system, including prompt washing after pesticide use, help prolong its lifespan and prevent cross-contamination of chemicals. Ending each day with an empty tank and flushing the system with clean water further ensures proper equipment hygiene and performance.

Completing Spraying Operations

Completing crop boom spraying operations involves several steps to ensure effective application and safety. After completing the spraying tasks, the equipment needs to be properly packed up and stored. Here's a detailed explanation of both processes:

Completing Crop Boom Spraying Operations:

1. Assessment and Monitoring: Before starting, assess the field conditions and monitor weather forecasts to ensure optimal spraying conditions, avoiding high winds, rain, or extreme temperatures.

2. Calibration: Calibrate the sprayer to ensure accurate application rates based on the specific requirements of the pesticide or fertilizer being applied.

3. Field Setup: Position the tractor and sprayer at the edge of the field, ensuring safe access and minimizing crop damage during the spraying process.

4. Sprayer Setup: Check the sprayer components, including nozzles, hoses, and pumps, to ensure they are clean, properly connected, and functioning correctly.

5. Spraying Operations: Begin spraying, ensuring consistent coverage across the field while adhering to recommended application rates and safety guidelines. Maintain a steady speed and proper boom height to achieve uniform application.

6. Monitoring and Adjustments: Continuously monitor spray patterns and adjust equipment settings as needed to address any issues such as clogged nozzles or uneven application.

7. Record Keeping: Keep detailed records of spraying activities, including the type and amount of chemicals used, weather conditions, and any observations related to crop health or equipment performance.

Packing Up Equipment:

1. Cleaning: After completing spraying operations, thoroughly

clean all equipment components, including the tank, nozzles, hoses, and filters, to remove any residue or chemical buildup. Use appropriate cleaning agents and follow manufacturer guidelines for cleaning procedures.

2. Drainage: Drain any remaining liquid from the tank and hoses, ensuring proper disposal of excess chemicals according to regulations.

3. Rinsing: Rinse the sprayer system with clean water multiple times to flush out any remaining traces of chemicals, preventing cross-contamination during storage.

4. Dismantling: If necessary, dismantle and disassemble components of the sprayer for more thorough cleaning and inspection, paying close attention to areas prone to chemical buildup or corrosion.

5. Storage: Store all cleaned and dried equipment in a secure, well-ventilated location away from direct sunlight and extreme temperatures. Ensure proper storage of chemicals according to safety guidelines and regulations.

6. Maintenance: Perform routine maintenance tasks such as lubrication, inspection of seals and fittings, and replacement of worn parts to keep the equipment in good working condition and prolong its lifespan.

CHAPTER EIGHT

Combine Harvester

A combine harvester, often simply called a combine, is a versatile agricultural machine designed to efficiently harvest a variety of grain crops such as wheat, barley, corn, oats, and soybeans. It combines several harvesting tasks into a single operation, including reaping, threshing, and winnowing, which traditionally required separate machines and considerable manual labour.

The term "combine harvester" originates from its ability to amalgamate three distinct functions: reaping, threshing, and winnowing. Reaping involves cutting and gathering the crops, while threshing loosens the grain from the dry casing surrounding the seed. Winnowing further refines this process by separating the grain from the scaly plant material, effectively separating the valuable grain from the unwanted chaff.

Here's a breakdown of the main components and functions of a combine harvester:

1. Header: The header is the front part of the combine that contains the cutting mechanism. It consists of a reel, cutter bar, and sometimes a draper belt or auger to feed the crop into the machine.

2. Threshing and Separation: As the crop is cut by the header and fed into the combine, it passes through a rotating cylinder or rotor equipped with threshing elements (such as teeth or concaves) that separate the grain kernels from the stalks and husks. The separated grain falls through sieves or screens, while the remaining straw and chaff are expelled out the back of the machine.

3. Cleaning: The separated grain undergoes further cleaning to remove any remaining debris, such as chaff, dust, or small weed seeds. This is typically done using a combination of sieves, fans, and air blasts to separate the grain from lighter materials.

4. Grain Storage: The cleaned grain is collected and stored in a grain tank located on the combine. Modern combines have large storage capacities, allowing for continuous harvesting without frequent unloading.

5. Unloading System: When the grain tank is full, the combine can unload the harvested grain into a waiting truck or trailer using an unloading auger or conveyor system.

6. Control Systems: Many modern combines are equipped with advanced computerized control systems that monitor and optimize various aspects of the harvesting process, including crop flow, engine performance, and grain quality.

Combines come in a range of sizes and configurations to suit different farm sizes, crop types, and harvesting conditions. They have revolutionized the agricultural industry by significantly increasing the efficiency and productivity of grain harvesting operations, reducing labour requirements, and allowing farmers to harvest large areas of crops in a relatively short amount of time.

Figure 56: A John Deere combine harvester harvesting the barley at Tower Farm. A John Deere combine harvester. Harvesting the barley at Tower Farm. Evelyn Simak / A combine harvester, CC BY-SA 2.0, via Wikimedia Commons.

Modern combine harvesters epitomize the fusion of cutting-edge technology with agricultural expertise, representing a remarkable synergy that revolutionizes the harvesting process.

Breaking Down the Components of a Combine Harvester — To gain insight into the inner workings of a combine harvester, it's essential to dissect its various components.

Power Source and Engine — At the heart of every combine lies a powerful engine that serves as the driving force behind the entire operation. These engines are meticulously engineered for efficiency and reliability, ensuring uninterrupted performance during long hours of fieldwork.

Threshing Mechanism — The threshing mechanism is tasked with the crucial job of separating the grain from the harvested crop. It employs a combination of rotating cylinders and concaves to achieve this fundamental function with precision.

Separation System — Once the threshing process is complete, the separation system further refines the harvested material, meticulously separating the grain from the chaff and straw, ensuring the purity of the yield.

Grain Tank — Serving as the repository for the harvested grain, a capacious tank onboard the combine provides ample storage capacity, minimizing the need for frequent unloading and enhancing operational efficiency.

Straw Chopper — Concluding the harvesting cycle, the straw chopper finely chops leftover straw and chaff into smaller pieces. These remnants can then be dispersed back onto the field as valuable mulch, contributing to soil health and fertility.

While combine harvesters often receive considerable attention, the header parts play an equally crucial role in the harvesting process, albeit without much recognition. These essential components ensure the efficient feeding of the crop into the machine for processing.

Header Platforms — Available in various sizes tailored to different crop types, header platforms are responsible for cutting and gathering the crop.

Reel and Knife System — The reel and knife system serves a pivotal function in directing the crop into the header, ensuring a clean and effective cut.

Auger and Conveyor Belts — Following the cutting process, augers and conveyor belts transport the crop to the threshing mechanism of the combine harvester.

Draper Belts — A prevalent feature in modern header designs, draper belts offer enhanced flexibility and versatility when harvesting diverse crops.

Two primary types of traction are utilized for combine harvesters: self-propelled and tractor-mounted.

Self-propelled wheeled harvester machines are particularly suitable for farms with compacted soil conditions, making them the predominant choice in agriculture.

Tractor-mounted harvesters, on the other hand, are affixed to and towed by a tractor. They are most effective in regions with loose soil and vast farming expanses.

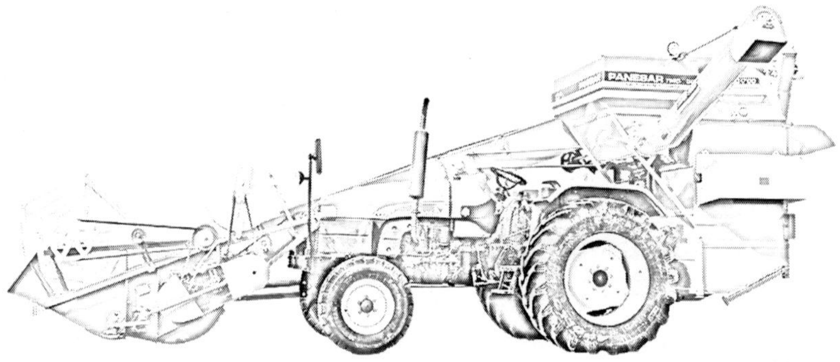

Figure 57: Tractor-mounted combine harvester arrangement.

In certain instances, combine harvesters are equipped with tracks instead of wheels, enhancing efficiency in areas prone to wheel sinking. Track-equipped combines are frequently deployed for harvesting rice and other crops in marshy or wetland environments.

There are three varieties of combine harvesters designed for different methods of grain and chaff separation: those equipped with shakers (also known as straw walkers), those with rotors, and hybrid models. Your selection among these options will be influenced by the specific requirements of your harvesting area and your objectives.

The Conventional Combine Harvester (with shakers): This traditional choice has been favoured by farmers for its reliability and long-standing technology. It typically features 4 to 8 shakers for effective grain and chaff separation.

The Non-conventional Combine Harvester (with rotors): This type of harvester boasts a more compact design compared to shaker-equipped models, making it less cumbersome. It also demands less maintenance. The resulting product tends to be of higher quality due to reduced grain breakage. However, it is sensitive to factors such as the mass and condition of the straw, as well as its moisture content. Additionally, it typically exhibits higher fuel consumption per hour.

The Hybrid Combine Harvester: Combining elements of both shaker and rotor technology, this hybrid machine incorporates a drum and concave alongside rotors. As a result, threshing operations result in less damage to the grain. However, hybrid models are less readily available on the market compared to their counterparts.

The main difference between conventional and non-conventional combine harvesters lies in their method of separating grain and chaff during the harvesting process.

1. **Conventional Combine Harvesters:**

    - **Separation Method:** Conventional combine harvesters use a shaking mechanism, often referred to as straw walkers, to separate the grain from the chaff. The harvested material passes through a series of oscillating sieves or shakers, allowing the grain to fall through while retaining the straw and chaff.

    - **Design:** These harvesters typically feature a design with straw walkers, which have been a longstanding and reliable technology in agriculture.

    - **Reliability:** They are known for their reliability and are the traditional choice for many farmers due to their proven track record.

- **Number of Shakers:** The number of shakers can vary, typically ranging from 4 to 8, depending on the model and manufacturer.

2. **Non-Conventional Combine Harvesters:**

   - **Separation Method:** Non-conventional combine harvesters utilize rotor technology for grain separation instead of the traditional shaking mechanism. Rotors are cylindrical devices that spin rapidly to separate the grain from the straw and chaff by centrifugal force.

   - **Compact Design:** These harvesters often feature a more compact design compared to conventional ones, making them less cumbersome and easier to manoeuvre in the field.

   - **Less Maintenance:** They generally require less maintenance compared to conventional models, offering improved efficiency in terms of upkeep.

   - **Grain Quality:** Non-conventional harvesters are known to produce grain of better quality as the threshing process is gentler, resulting in less grain breakage.

   - **Sensitivity to Conditions:** However, they can be more sensitive to factors such as the mass and condition of the straw, as well as humidity levels, which can affect their performance.

   - **Fuel Consumption:** Non-conventional harvesters may have higher fuel consumption rates per hour compared to conventional models due to the rotor technology and other design features.

In summary, conventional combine harvesters rely on shaking mechanisms for grain separation and are known for their reliability, while non-conventional harvesters use rotor technology for separation, offering benefits such as improved grain quality and reduced maintenance requirements, albeit with potential sensitivity to certain conditions and higher fuel consumption.

Choosing the appropriate combine harvester depends on the size of the harvest area and specific requirements.

For smaller areas under 300 hectares, a single conventional combine harvester with straw walkers is often sufficient for efficient harvesting.

In cases where the area exceeds 300 hectares or where continuous harvesting is challenging, there are two options:

1. Utilizing two conventional combines can enhance productivity and manageability, albeit requiring additional labour and resources.

2. Opting for a non-conventional combine harvester equipped with a rotor can offer greater capacity and efficiency while avoiding the need for daily maintenance of multiple machines.

Determining the storage capacity of the combine harvester is crucial to ensure optimal performance during harvesting. The storage capacity of the hopper should align with the processing capabilities of the harvesting head, with various options ranging from 4,000 to 10,000 litres. Selecting an appropriate storage capacity promotes uninterrupted harvesting operations.

Consideration should also be given to the choice of tyres for the combine harvester. Standard combine harvesters weigh around 15 tonnes on average, which, when combined with a filled hopper, can exceed 30 tonnes. This substantial weight poses the risk of soil compaction, potentially impacting productivity in subsequent harvests. Tailoring the tyres to the combine harvester's storage capacity and terrain characteristics can mitigate soil compaction issues. Alternatively, opting for a combine harvester equipped with tracks offers an effective solution to reduce soil compaction and enhance manoeuvrability in diverse field conditions.

A typical combine harvester works through a series of mechanical processes designed to efficiently harvest crops. Here's a simplified explanation of its operation:

1. Heading into the Field: The combine harvester is driven into the field where the crops are ready for harvesting.

2. Cutting: The harvesting process begins with the cutting mechanism, typically a header at the front of the machine, which cuts the crop stalks. The header may be equipped with various attachments depending on the crop type and field conditions.

3. Feeding: Once cut, the crop material is fed into the machine through the header. The feeder mechanism guides the crop into the threshing unit.

4. Threshing: Inside the combine, the threshing unit separates the grain (such as wheat, corn, or rice) from the stalks and husks. This is achieved by rotating drums or rotors, which beat the crop material to release the grain.

5. Separation: After threshing, the separated grain is separated from the remaining straw, chaff, and husks. This is typically done using a combination of sieves and airflow, which allow the lighter material to be blown away while the heavier grain falls into a collection tank.

6. Cleaning: The separated grain is then cleaned to remove any remaining debris or impurities. This is usually accomplished using screens, air currents, and gravity to ensure that only clean

grain is collected.

7. Storage: The cleaned grain is stored in a grain tank onboard the combine until it is ready to be unloaded. Modern combines often have large storage capacities to minimize the need for frequent unloading.

8. Discharging: Once the grain tank is full or the harvesting process is complete, the grain is discharged from the combine either directly into a waiting truck or into storage bins.

9. Processing Residue: The remaining straw, chaff, and husks, known as crop residue, are typically ejected from the rear of the combine. In some cases, this residue may be chopped into smaller pieces and spread back onto the field as mulch.

10. Continuous Operation: Throughout the harvesting process, the combine is continuously driven through the field in a methodical pattern to ensure that all the crops are efficiently harvested.

Overall, a combine harvester streamlines the harvesting process by integrating cutting, threshing, separating, and cleaning functions into a single machine, thereby increasing efficiency and productivity in agricultural operations. Figure 58 and Figure 59 show the mechanisms and the flow of the process.

Figure 58: Threshing mechanisms. Back image - kallerna, CC BY-SA 4.0, via Wikimedia Commons.

Mechanical components commonly utilized in a combine harvester are as follows:

Reel: The reel, an essential rotating component, plays a pivotal role in lifting the crop from the ground and guiding it towards the cutter bar. It consists of several integral parts, including reel bats, which are metal arms extending from the reel shaft and rotating to lift the crop, and fingers, smaller metal tines attached to the reel bats that aid in lifting the crop. Additionally, bearings support the reel shaft, ensuring smooth rotation, while the drive mechanism propels the reel shaft to rotate. Operators can control the speed of the reel rotation and adjust its height to suit the crop height.

Cutter Bar: A crucial component of a combine harvester, the cutter bar facilitates the cutting of the crop as the harvester moves through the field. Comprising knife sections mounted on the bar for cutting, guards to shield the knife sections from damage, hold-downs to secure the crop in place during cutting, and reel fingers similar to those found on the reel, the cutter bar ensures efficient crop harvesting. The knife drive mechanism propels the knife sections for cutting, while operators can control the speed of the knife sections' movement.

- Threshing Drum: Responsible for separating grain from straw and chaff, the threshing drum is a cylindrical component that rotates and contains bars, teeth, or rasp bars. These components rub against the crop as it passes through the drum, facilitating the separation process. Drum bars mounted around the circumference of the drum aid in threshing, while concaves mounted inside help further separate the grain from the straw and chaff. Rasp bars with teeth or projections assist in the separation process. Drum bearings support the threshing drum, allowing smooth rotation.

- Unloading Auger: The unloading auger, a mechanical arm extending from the grain tank, unloads harvested grain into a trailer or storage bin. It comprises several parts, including an auger tube, a long metal tube extending from the grain tank to the unloading point, auger flighting, a spiral-shaped metal blade within the tube that moves grain towards the unloading point, and flighting bearings supporting the auger flighting. Additionally, an unloading spout directs the grain into the trailer or storage bin, with an optional extension available to increase reach. A positioner mechanism allows operators to adjust the spout's position for precise grain deposition.

- Transmission and Engine: Among the most crucial components, the transmission and engine power and control the combine harvester. The transmission housing encloses gears and other transmission components, while gears transmit power from the engine to the wheels and other moving parts. Shafts connect gears and other transmission elements. A clutch engages and disengages the transmission from the engine, enabling speed and direction control. The hydraulic system provides pressure and fluid to control the clutch and other transmission components.

These components work synergistically to power and operate the combine harvester efficiently during crop harvesting and processing. Proper maintenance and adjustment of the transmission and engine are imperative for ensuring smooth and reliable harvester operation.

While machines may vary in design, they generally follow the same fundamental principles, and a typical combine operates through several distinct stages. Initially, as the combine moves forward, the header gathers the crop. The rotating pick-up reel at the front draws the crop over the cutting blades.

Subsequently, the cut crops are lifted into the threshing cylinder or drum, where the grain is separated from the stalk through agitation. The separated grain is then sifted and collected in a storage tank.

Once the tank reaches its capacity, an elevator carries the grain upward through a side pipe or unloader and into a trailer positioned beside the combine. Meanwhile, a fan blows the chaff out of the sieves, while the remaining straw can be gathered and baled.

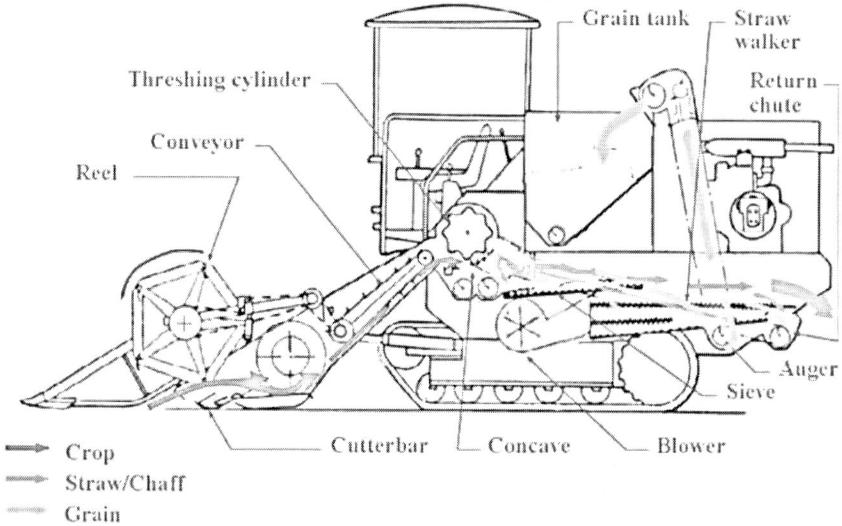

Figure 59: cross-sectional diagram of a typical combine harvester shows the path followed by a generic grain crop as it is harvested from a field. Adapted from International Rice Research Institute (IRRI), CC BY 3.0, via Wikimedia Commons.

While combine harvesters may appear uniform to many observers, there are several variations tailored to different crops and handling requirements. Most combines feature interchangeable headers, allowing them to harvest a wide range of crops.

For crops like cotton, specialized handling is necessary. For instance, the John Deere cotton harvester combines features of a traditional combine harvester at the front with those of a hay baler at the rear. Instead of dispersing grain and chaff, this machine compresses and packages the cotton into large round bales known as "modules," which are then wrapped in protective plastic. Once a bale is formed, the machine's hydraulic mechanism lifts it and deposits it onto the field behind.

Figure 60: The Case IH Module Express™ 625 (a combined cotton harvester and module builder) picks cotton efficiently and simultaneously builds ginner-friendly cotton modules reducing a cotton producer's equipment and labour investment while streamlining the harvest process. Cgoodwin, CC BY-SA 3.0, via Wikimedia Commons.

In the case of large headers, robust wheels are essential to support their weight, though this is not always the case. Some combines utilize half-tracks at the front, resembling miniature tanks. Although mechanically more intricate and costly, these half-tracks offer several advantages. They eliminate the need for oversized front wheels, allowing combines equipped with tracks to navigate narrower roads.

Additionally, they enhance stability when operating with wide headers and provide superior traction on hills and challenging terrain.

Combine headers play a crucial role in the process of combined harvesting, serving as the frontline attachment responsible for cutting and gathering grains and other crops. These headers are affixed to the front end of harvesting machinery and come in various shapes and sizes, depending on factors such as the type of crop, scale of production, and specific requirements of the farm. Choosing the appropriate combine header is paramount as an incorrect selection can adversely affect both the quality and quantity of grain yield. Thus, understanding the different types of combine headers and their respective applications is essential. The following are six major types of combine headers:

1. Grain Headers: Designed for cutting crops and collecting grains, grain headers utilize a canvas-style rolling platform belt to transport the crop from the sickle bar to the combine for further cleaning and grain separation. They are particularly suitable for cereal crops with small grains such as oats, peas, beans, wheat, and various other grains.

2. Corn Headers: Equipped with a spike structure and pull chain, corn headers are specialized for harvesting crops like corn and soybeans. These headers collect the ears while chopping the stalks, enhancing efficiency during corn harvesting.

3. Sunflower Headers: Specifically tailored for sunflower harvesting, these headers feature trays fitted to the cutting bars at the front. Head snatchers can be employed to facilitate fast and efficient harvesting, optimizing grain samples for maximum productivity by cutting stalks beneath the flower heads.

4. Pick-Up Headers: Also known as rake-up headers, pick-up headers utilize long rakes or tines to collect crops, which are then fed to the auger via belts. They are suitable for harvesting

lentils, beans, canola, and grass seeds by gathering crops directly from the windrow.

5. Flex Platform Headers: These headers are designed to adapt to uneven terrain, featuring knives and guards to cut crops effectively without picking up rocks or other ground materials. Farmers often combine flex and draper headers for enhanced results, making them suitable for various crops, including rice, especially in rough and rocky terrain.

6. Draper Headers: Draper headers utilize a conveyor belt system to convey crops from both ends towards the centre, where a third draper pushes them into the feeder house. Unlike conventional headers with augers, draper headers are equipped with conveyor belts to fetch grains, making them suitable for crops like wheat, barley, oats, rye, triticale, rice, and soybeans.

Planning and Preparing for Combine Harvester Operations

Combine harvester operations involve a series of essential steps to ensure safety, efficiency, and effectiveness. The first step is to determine work requirements from workplace instructions. This involves carefully reviewing instructions provided by supervisors or managers, which outline the specific tasks and objectives for the operation. Understanding the scope of work, the area to be harvested, and any special instructions is crucial for planning and executing the operation accurately.

Once the work requirements are established, the next step is to identify potential hazards and risks associated with combine harvester operations. This includes recognizing hazards such as moving machinery parts, uneven terrain, weather conditions, and proximity to other workers or objects. Implementing safe working practices is essential to manage these risks effectively. This may involve measures such as proper training, adherence to safety protocols, and the use of safety equipment like seat belts, roll bars, and personal protective equipment (PPE).

To ensure the safe and efficient operation of the combine harvester, it is crucial to source and interpret relevant machinery and equipment operation and maintenance manuals, as well as manufacturer instructions. These documents provide valuable information on operating procedures, maintenance requirements, and safety guidelines. Familiarizing oneself with this information helps in understanding the functions and capabilities of the equipment and ensures compliance with recommended practices.

Selecting, fitting, using, and maintaining personal protective equipment (PPE) applicable to the task is another critical aspect of combine harvester operations. Depending on the specific hazards present, appropriate PPE such as protective clothing, gloves, goggles, and ear protection should be worn to mitigate the risk of injury. Regular inspection and maintenance of PPE are essential to ensure its effectiveness and reliability during operation.

Before commencing harvesting activities, it is essential to select the required tools, equipment, and machinery and conduct routine pre-operational checks. This includes inspecting components such as the engine, cutting blades, hydraulic systems, and safety features to ensure they are in proper working condition. Any damaged or worn components should be documented and replaced according to workplace procedures to prevent equipment failure or accidents during operation.

Finally, attaching ancillary equipment and checking for correct operation is the last step before beginning harvesting activities. This involves attaching any additional equipment required for the specific task, such as header extensions or crop dividers, and verifying that all attachments are securely fastened and functioning correctly. Conducting a final check ensures the proper operation of all equipment and machinery, contributing to a safe and successful combine harvester operation.

For comprehensive information on adjusting a combine, it's imperative to refer to the operator's manual tailored to the specific model in use. To complement this manual, the following adjustment principles and recommendations are provided to reinforce and enhance understanding.

Fine-tuning the header height plays a crucial role during wheat harvest, significantly influencing various aspects such as harvest loss, combine capacity, operational expenses, and subsequent crop yields. Typically, setting the cutter bar height at the maximum level feasible without overlooking a few of the lowest heads is advisable. Head height uniformity varies based on factors like variety and growing conditions. Research conducted in Fort Collins, Colorado over three years indicated that a cutting height set at approximately 2/ the average head height resulted in minimal loss, below 0.5 percent. Lower heads generally contain less grain than the average (Smith et al., 2024).

In wheat harvesting, material other than grain (MOG) often governs separator loss. Hence, taller stubble enhances combine capacity and reduces fuel consumption per acre. Elevating the cutting height can yield fuel savings of up to 30 percent while adhering to acceptable loss thresholds. Moreover, taller stubble mitigates straw-related challenges for subsequent cropping systems, especially when followed by immediate double cropping after wheat harvest.

Figure 61: Harvesting wheat. Harvesting wheat by Philip Halling, CC BY-SA 2.0, via Wikimedia Commons.

Taller stubble also curbs evaporation, with an increase in stubble height from 4 to 20 inches reducing potential evaporation by around 40 percent, at a stem density of 170 stems per square yard. Elevating wheat stubble height from 4 inches to 10 inches can substantially elevate subsequent no-till corn yield, from 40 to nearly 65 bushels per acre. Additionally, corn and grain sorghum yields may witness a five-bushel-per-acre boost following stripped wheat. Furthermore, taller stubble augments wildlife habitat, with an elevation from 9 inches to 18 inches resulting in a nearly nine-fold increase in winter pheasant populations in western High Plains wheat fields.

Adjusting travel speed is a common operator manoeuvre. In conventional cylinder combines, straw walkers are often the first to succumb to overload. This becomes apparent only with the correct use of a loss monitor. Given the considerable spatial variation in grain yields, frequent adjustments in combine travel speed are necessary for optimal performance. Generally, rotary combines exhibit lower susceptibility to separator overload stemming from excessive speed or MOG input (Smith et al., 2024).

Fine-tune table auger finger timing to ensure smooth feeding into the feeder house. During the harvesting of light, droughty wheat, adjust fingers to extend later. Typically, finger timing adjustment entails rotating a plate located on the undriven end of the table auger, as detailed in the owner's manual.

Set table auger strippers as close as practical to the auger flighting, taking care to allow clearance for auger run-out, including transient thermal warping due to direct sunlight exposure. Wider headers necessitate increased clearance between auger and strippers. Floor strippers are recommended where available.

Optimize cylinder or rotor speed and clearance to thresh the crop minimally required for efficient grain dislodgement and separation. Over-threshing not only squanders power but can also lead to grain cracking, cleaning shoe overload, baling impede, and accelerated wear of concaves, bars, and drive systems. Cracked grain typically results from excessive cylinder speed rather than insufficient cylinder-concave clearance.

Most combines prevalent in the Great Plains are "corn-soybean" models, as the penalty for employing such a combine in wheat is lower than the converse. However, wheat growers might find it beneficial to substitute the 1% inch louvered corn-bean chaffer with a 1 1/8 inch wheat chaffer. Additionally, fixed airfoil and adjustable Peterson chaffers are available for most machines and warrant consideration, particularly if the combine is also utilized in canola.

While winter annual grasses are ideally managed through herbicides and crop rotations, certain practices are recommended for harvesting infested crops (Smith et al., 2024):

1. Prioritize harvesting heavily infested wheat fields last to afford weedy grasses ample drying time.

2. Adjust the chaffer towards the open end of the recommended range.

3. Position the cleaning sieve towards the closed end of the recommended range.

4. Adjust the fan towards the high end of its recommended range.

5. Vigilantly monitor return rates and make necessary adjustments.

6. Strategically manage combine traffic patterns to prevent the potential spread of weed seeds from isolated infestations over a wide area.

Prepare to accept a minor moisture dock at the outset of harvest. Some farmers delay wheat harvest until it attains a moisture level acceptable for commercial channels with zero moisture dockage. However, prolonged wheat harvest durations may yield overly dry (less than 10 percent) wheat towards the end. Overly dry wheat translates to lost income akin to moisture dockage that would have been incurred with an earlier start. Opting for a balanced approach, accepting a modest moisture dock at the harvest's onset, results in earlier completion, reduced weather exposure, and enhanced double-cropping potential.

Combines and grain carts rank among the heaviest equipment traversing fields, potentially inducing significant soil compaction if not equipped with appropriately designed tyre or track systems. Large combines boasting wide headers and 300-plus-bushel grain tanks may approach loaded weights nearing 60,000 pounds. Grain carts, too, have expanded in size, with capacities reaching 1400 bushels and loaded weights nearing 100,000 pounds. However, soil compaction is not solely dictated by implement weight; various factors influence its occurrence.

As soil moisture content rises, the soil's resistance to mechanical deformation decreases, heightening the likelihood of soil compaction. Ideally, refrain from field operations when the soil is excessively wet.

Soil compaction accumulates with repeated passes of tractors or implements. While a single pass might induce only low or moderate compaction, subsequent passes exacerbate the issue. Endeavor to minimize repeat traffic or designate specific routes to limit compaction zones.

Soil compaction hinges not only on implement weight but also on the pressure exerted by the tyre or track surface on the soil. Managing this contact pressure is pivotal in averting soil compaction, necessitating the use of tyres or tracks capable of adequately distributing the load.

Soil compaction risks can be mitigated through a combination of management strategies (Smith et al., 2024):

- Avoiding wet fields whenever feasible and allowing sufficient drying time.
- Minimizing repeat traffic or establishing designated traffic zones.
- Limiting tyre or track-to-soil contact pressure to below 10 psi, or at least under 15 psi.
- Opting for wide belted tracks or large radial tyres with low inflation pressure to ensure adequate floatation.
- When soil conditions suggest compaction risks, limit combine grain tank or grain cart loading to half capacity to alleviate axle weight and tyre or track-to-soil pressure.
- Strategically positioning trucks, providing multiple access

points, and unloading prematurely near the truck to minimize combine and cart travel.

- Utilizing local band radios to coordinate grain cart operations efficiently, reducing unnecessary travel for both grain carts and combines during unloading operations.

As an example of process, harvesting wheat poses significant challenges that necessitate thorough preparation and precise timing. Leaving dry wheat in the field for too long exposes it to the risk of damage from winds and storms, potentially compromising the crop's quality. Additionally, if wheat is rained on and then dries again, its quality may further deteriorate. Moreover, the harvesting process itself requires the use of a combine, a hefty machine that demands training and careful handling. While maintenance and operation of a combine can be managed by a single individual, large-scale wheat harvests typically involve a team operating multiple combines and trucks.

Preparing to harvest wheat includes:

1. Assess Wheat Moisture Level:
    - Utilize a moisture meter to gauge the wheat's moisture content, crucial for determining its readiness for harvest.
    - Optimal harvesting occurs when the grain's moisture content ranges between 20% and 14%.
    - Moisture meters are readily available at agricultural and farming supply stores.
2. Conduct Combine Maintenance:
    - Ensure peak performance by performing necessary maintenance tasks on the combine, referring to the owner's manual for specific requirements.

- Check the sharpness of the sickle for optimal cutting performance.
- Verify the settings of height and contour controls on the headers.
- Follow the manual's guidelines to grease all components adequately for smooth operation.

3. Inspect Feeder House:
    - Thoroughly examine the feeder house of the combine to confirm its proper functioning.
    - Regular maintenance prevents potential breakdowns despite the feeder house's apparent durability.
    - Replace any broken, bent, or worn slats and chains.
    - Assess the condition of the drive belt, replacing it if any signs of damage such as cracking are observed.
4. Routine Equipment Inspection:
    - Cultivate a habit of inspecting equipment before each use to mitigate the risk of overlooking critical issues.
    - Monitor the air pressure in the combine's tyres on a weekly basis.
    - Ensure the combine is adequately fuelled before embarking on harvesting tasks.
    - Regularly check oil and radiator levels to maintain optimal engine performance.
    - Thoroughly clean the machine to remove dust, debris, and

other potential operational impediments.

- Verify the functionality of lights and flashers, especially if traveling on public roads during harvesting activities.

Combine harvesters, while essential for modern farming practices, pose several hazards to operators and bystanders alike. Some of the main hazards associated with combine harvesters include:

1. Entanglement: Moving parts such as the cutter bar, reel, and augers can catch loose clothing, hair, or body parts, leading to severe injuries or even death if an individual becomes entangled in the machinery.

2. Crushing and Pinching: The sheer size and weight of combine harvesters make them capable of causing crushing and pinching injuries if individuals come into contact with moving or stationary parts such as wheels, belts, or hydraulic components.

3. Falls: Operators working on or around the combine harvester, particularly when performing maintenance or adjustments, risk falling from elevated platforms or slipping on surfaces made slippery by oil, fuel, or crop residue.

4. Electrocution: Electrical components of the combine harvester pose a risk of electrocution, especially if damaged wiring or components come into contact with moisture or conductive materials.

5. Fire and Explosion: The presence of flammable materials such as crop residue, fuel, and hydraulic fluid increases the risk of fire or explosion, particularly if hot engine components come into contact with combustible materials.

6. Chemical Exposure: Pesticides, herbicides, and other chemicals

used in crop management may be present on the crop or within the machinery, posing a risk of exposure to operators and maintenance personnel.

7. Noise: Continuous operation of combine harvesters generates high levels of noise, potentially causing hearing damage or impairment to operators and bystanders if proper hearing protection is not worn.

8. Visibility Issues: Limited visibility from the operator's cab can lead to accidents involving bystanders, particularly in crowded or obstructed work areas such as fields with uneven terrain or tall crops.

9. Fatigue and Ergonomic Hazards: Prolonged periods of operation can lead to operator fatigue, increasing the risk of accidents due to reduced concentration and impaired reaction times. Additionally, poor ergonomic design of controls and seating may contribute to musculoskeletal disorders.

10. Transportation Risks: Moving combine harvesters between fields or on public roads presents risks of collisions with other vehicles, particularly due to their large size and slower speeds compared to regular traffic.

It is essential for operators and workers to undergo thorough training on safe operation and maintenance practices, adhere to manufacturer guidelines and safety protocols, and use appropriate personal protective equipment to mitigate these hazards effectively. Regular maintenance and inspection of machinery are also crucial for identifying and addressing potential safety issues before they escalate into accidents.

Mitigation strategies for the hazards associated with combine harvesters are essential to ensure the safety of operators and bystanders. These can include:

1. **Entanglement**:
   - Install guards and shields over moving parts to prevent contact with individuals.
   - Train operators on safe clothing practices and prohibit loose-fitting clothing, jewellery, or long hair while operating the machinery.
   - Implement emergency stop devices that quickly shut off power to moving parts in case of entanglement.
2. **Crushing and Pinching**:
   - Establish clear exclusion zones around the combine harvester to prevent unauthorized personnel from entering hazardous areas.
   - Ensure proper maintenance of brakes and hydraulic systems to prevent unintended movement.
   - Provide training to operators on safe operating practices and the risks associated with pinch points.
3. **Falls**:
   - Use anti-slip surfaces and handholds on elevated platforms to prevent slips and falls.
   - Require operators to wear fall protection equipment, such as harnesses and lanyards, when working at heights.
   - Conduct regular inspections of platforms and access points

for structural integrity.

4. **Electrocution**:
    - Inspect electrical components regularly for damage and replace worn or frayed wiring immediately.
    - Ensure that electrical systems are properly grounded and insulated to prevent shocks.
    - Provide training to operators on the safe handling of electrical components and the identification of electrical hazards.
5. **Fire and Explosion**:
    - Maintain a clean and debris-free engine compartment to reduce the risk of ignition sources.
    - Store flammable materials, such as fuel and hydraulic fluid, in designated areas away from heat sources.
    - Equip the combine harvester with fire extinguishers and train operators on their proper use in case of fire.
6. **Chemical Exposure**:
    - Wear appropriate personal protective equipment, such as gloves, goggles, and respiratory protection, when handling chemicals.
    - Implement proper handling and disposal procedures for pesticides and herbicides to minimize exposure.
    - Provide training to operators on the safe use and handling of agricultural chemicals.
7. **Noise**:

- Use noise-reducing engineering controls, such as sound-insulated cabs and mufflers, to reduce exposure to high noise levels.
- Require operators to wear hearing protection, such as earplugs or earmuffs, when operating the machinery for extended periods.

8. **Visibility Issues**:
    - Install mirrors and cameras on the combine harvester to improve visibility around the machine.
    - Use spotters or flaggers to assist operators when manoeuvring in crowded or obstructed areas.
9. **Fatigue and Ergonomic Hazards**:
    - Schedule regular breaks for operators to prevent fatigue and allow for rest and recovery.
    - Provide ergonomic seating and controls to reduce the risk of musculoskeletal injuries.
    - Rotate tasks to vary the physical demands on operators and reduce the risk of overexertion.
10. **Transportation Risks**:
    - Follow safe transportation practices, such as using escort vehicles and proper signage, when moving combine harvesters on public roads.
    - Ensure that all lights and indicators are functioning properly to improve visibility to other road users.

- Provide training to operators on safe driving practices and the unique handling characteristics of combine harvesters on public roads.

Operating a Combine Harvester

Carrying out pre-start and start-up procedures according to workplace procedures is a crucial aspect of combine harvester operations. Before starting the machinery, it is essential to conduct a thorough pre-start inspection to ensure that all components are in proper working order. This includes checking the engine oil level, hydraulic fluid level, fuel level, and inspecting the tyres or tracks for any signs of damage or wear. Additionally, inspecting safety features such as seat belts, roll bars, and emergency stop buttons ensures that the equipment is safe for operation. Following workplace procedures during the pre-start phase helps in preventing potential accidents and ensures the smooth start-up of the combine harvester.

Conducting a prestart check on a combine harvester is essential to ensure that the machine is in proper working condition and safe to operate. Here's a comprehensive guide on how to conduct a prestart check:

1. **Review the Operator's Manual:** Start by familiarizing yourself with the operator's manual provided by the manufacturer. Pay attention to any specific prestart checklists or procedures outlined in the manual.

2. **Visual Inspection:** Perform a thorough visual inspection of the entire combine harvester, checking for signs of damage, wear, or loose parts. Inspect the exterior and interior components, including critical areas such as the engine compartment, wheels,

and harvesting mechanisms.

3. **Fluid Levels:** Check the fluid levels of essential components such as engine oil, hydraulic fluid, coolant, and fuel. Ensure that all fluid levels are within the recommended range as specified in the operator's manual. Top up any fluids that are low and address any leaks.

4. **Tyre Inspection:** Inspect the tyres for proper inflation, tread wear, and damage. Ensure that all tyres are properly inflated according to the manufacturer's recommendations. Look for any signs of punctures, cuts, or bulges that may affect the machine's performance or safety.

5. **Electrical System:** Check the battery terminals for corrosion and ensure they are securely connected. Test the headlights, indicators, and other electrical components to ensure they function correctly. Inspect wiring harnesses for damage.

6. **Brake Inspection:** Test the brakes to ensure they function correctly. Check for proper brake pedal resistance and ensure the parking brake engages and disengages smoothly. If equipped with air brakes, check the air pressure levels.

7. **Safety Features:** Verify that all safety features and devices are in working order, including seat belts, emergency stop switches, warning lights, and audible alarms. Test each safety feature to ensure they function as intended.

8. **Functional Checks:** Start the engine and allow it to idle while observing for any abnormal noises or warning lights. Engage the harvesting mechanisms and verify they operate smoothly.

9. **Test Drive:** If possible, perform a brief test drive to assess

overall performance and handling. Pay attention to steering responsiveness, braking efficiency, and engine performance.

10. **Documentation:** Record the results of the prestart check in a maintenance log or checklist. Document any defects or abnormalities discovered and report them for further investigation.

Additionally, follow these steps to ensure the combine harvester is prepared for safe and efficient operation:

- Clear any dust, dirt, or debris from the machine.
- Inspect for loose wires, signs of wear and tear, and wheel condition.
- Attach headers and ensure they are in good working condition.
- Adjust height and contour controls for optimal operation.
- Check sickle blades, guards, augers, and reel teeth for damage.
- Inspect gathering chains, sprockets, and stripper plates.
- Lubricate row-unit boxes and inspect chains and bearings.
- Check clearance between rotating cylinder and combine concave.
- Inspect threshing and auger components for damage.
- Check and fill oil levels, including engine oil, fuel filter, air cleaner, and hydraulic oil.
- Inspect lights, flashers, and reflectors for safety.
- Grease all points and check for any fittings that do not take grease.

Consult the operator's manual for specific instructions on any parts or issues you are unsure about. By following these steps, you can ensure your combine harvester is properly inspected and ready for safe and efficient operation.

Once the pre-start procedures are completed, operators must operate the machinery and equipment in a safe, controlled, and efficient manner while continuously monitoring performance efficiency. This involves understanding and adhering to safe operating practices outlined in the operation and maintenance manual, manufacturer specifications, and task requirements. Operators should be vigilant in monitoring engine performance, hydraulic systems, cutting and threshing mechanisms, and other critical components to ensure optimal efficiency and productivity. Regular monitoring helps identify any issues or malfunctions promptly, allowing for timely intervention and minimizing downtime.

Operating the machinery according to operation and maintenance manuals, manufacturer specifications, task requirements, and prevailing conditions is essential for the safe and effective operation of the combine harvester. This includes adjusting settings such as cutting height, reel speed, and thresher control based on the crop type, density, and moisture content. Adhering to recommended operating procedures helps prevent damage to the machinery, reduces the risk of accidents, and ensures the quality of harvested crops.

Identifying environmental and biosecurity implications associated with harvesting is another important consideration during combine harvester operations. Operators should be mindful of potential environmental impacts such as soil compaction, crop residue management, and the spread of weeds or pests. Implementing measures to minimize these impacts, such as using controlled traffic farming techniques or adopting integrated pest management practices, helps mitigate environmental and biosecurity risks associated with harvesting activities.

Continually monitoring hazards and risks and ensuring the safety of self, other personnel, plant, and equipment is a fundamental responsibility of combine harvester operators. This involves remaining vigilant and proactive in identifying potential hazards and taking appropriate measures to mitigate risks. Regularly assessing the operating environment, maintaining clear communication with other personnel, and adhering to safety protocols contribute to a safe and secure working environment. By prioritizing safety and risk management throughout the harvesting process, operators can minimize accidents, protect personnel and equipment, and optimize overall operational efficiency.

The operation of a combine harvester involves several distinct stages, each dedicated to a specific task in the harvesting process.

Field Entry and Preparation: Before commencing harvesting, the operator ensures that the combine harvester is correctly configured and adjusted based on the crop type. This includes setting the cutting height, reel speed, and other parameters to optimize efficiency and minimize crop wastage.

Crop Cutting: Once the combine harvester is properly configured, the cutting mechanism initiates operation. The header cuts the crop close to the ground, while the reel gathers and feeds it towards the cutter bar. The rotating sickle sections of the cutter bar then sever the main stems of the crop, effectively detaching it from the soil.

Grain Separation: After cutting, the crop enters the threshing mechanism. Here, the thresher drum rotates swiftly, using friction and rotational motion to separate the grain from the remaining plant material. The concave and beater bars further aid in this separation process, ensuring a thorough extraction of grain.

Grain Cleaning: Following separation, the grain undergoes a cleaning process. Passing through the sieve system, impurities such as straw, chaff, and larger foreign particles are removed. A powerful fan generates airflow to blow away lighter materials, while the chaffer assists in further segregating the grain from impurities.

Grain Storage: The cleaned grain is deposited into the grain tank, which has a substantial capacity to accommodate harvested grain. Equipped with sensors, the tank monitors grain levels to prevent over-filling. The grain remains securely stored until it is ready for unloading.

Grain Unloading: Upon reaching full capacity or when unloading is necessary, the unloading system is activated. An auger or conveyor belt system transfers the grain from the tank to a designated storage container, such as a truck or grain cart. Once unloading is complete, the combine harvester is prepared to resume the harvesting operation.

Instructions for starting a combine harvester:

1. Familiarize yourself with the operator's manual of the combine harvester. Locate the levers, pedals, and other controls as their positions may vary among different brands and models.

2. Conduct a safety inspection of the vehicle. Ensure that the brakes are engaged, levers are in neutral position, the engine is turned off, and the ignition key is removed. Check the header and auger to ensure there are no obstructions. Only enter the cab and assume the driver's seat once these conditions are met.

3. Initiate the starting sequence for the combine harvester. Typically, this involves pressing down on the fuel pedal and turning the ignition. Increase the engine speed by advancing the relevant lever. Rev the engine while the combine remains in neutral.

4. Adjust the position of the header or cutter accordingly. Raise it higher for harvesting wet crops and lower for dry conditions.

5. Progressively advance the thresher control followed by the header lever. Maintain this sequence as it ensures that the grain detachment mechanism operates at full capacity before the crop

is drawn into the machine. When ceasing operations, reverse the order.

6. Disengage the brake and push the drive lever forward. Select the appropriate gear based on the type and condition of the crop being harvested. Refer to the operator's manual for guidance on the safe and effective turning radius specific to the combine model.

7. Apply the brake upon completion of harvesting. Return all controls to the neutral position. Clean the blades and cylinders when the engine is turned off and the ignition key is removed.

Ensure safe operation around combine equipment by adhering to safety instructions outlined in the operator's manual at all times. Prior to commencing harvest activities, meticulously follow the manufacturer's maintenance checklist and inspect the machinery for any signs of wear or damage. Focus on checking critical components such as roller chains on corn heads or knives on cutterbar heads for soybeans and small grains. Additionally, in the threshing and separation area, thoroughly examine the rotor and concave for any indications of wear or damage, and assess the condition of the sieves in the cleaning shoe area. It is imperative to inspect all augers for sharp edges, as they can potentially cause grain damage. Replace any worn parts promptly to ensure optimal efficiency during the harvest process.

The header of the combine plays a pivotal role in the harvesting process, serving as the initial point of contact with the crop and significantly impacting grain loss. Various components within the header, such as gathering chains, stripping rolls, and deck plates, must be meticulously adjusted to minimize grain loss. Proper spacing of stripping rolls according to stalk thickness and adjustment of deck plates to prevent ear and kernel loss are essential. Furthermore, gathering chains must be appropriately set to ensure efficient operation, with particular attention paid to matching feederhouse and corn head gathering speeds to the combine drive speed. Vigilance is crucial while harvesting, as adjustments may need to be made in response to changes in crop and field conditions.

The feederhouse, where the grain first enters the combine, requires precise adjustments and settings tailored to the specific crop being harvested. Proper positioning of the feed drum and adjustment of the feed accelerator are vital to prevent grain damage and loss. It is imperative to set the feed accelerator at a low speed to ensure intact ears are conveyed into the combine for the threshing process. Incorrect accelerator settings can result in broken corn cobs and kernel loss before threshing begins.

Threshing adjustments are critical for optimizing harvest efficiency, with careful consideration given to rotor speed and concave clearance. Striking a balance between these parameters is essential to avoid grain damage and achieve thorough threshing. It is recommended to start with factory-recommended settings and fine-tune as necessary based on crop and field conditions. Maintaining a full rotor chamber, minimizing rotor speed, closing concave spacing, and adjusting rotor speed conservatively are key guidelines for effective threshing and separation.

Following threshing, grain cleaning adjustments play a crucial role in separating grain from non-grain crop material. Proper settings for chaffer and shoe sieves, along with fan speed, are essential for delivering a clean, high-quality end product. Failure to optimize these settings can result in impurities in the grain tank or loss of grain out the back of the combine. Sieve adjustments should complement fan speed adjustments for optimal results, ensuring efficient grain cleaning and minimal grain loss.

Figure 62: Wheat harvesting in Turkey. Maurice Flesier, CC BY-SA 4.0, via Wikimedia Commons.

The controls in the cabin of a combine harvester are essential for the operator to efficiently and safely operate the machine. Here's an overview of typical controls found in the cabin:

1. Steering Wheel: Similar to a vehicle, the steering wheel allows the operator to control the direction of the combine harvester. It's usually located in front of the operator's seat and is used to navigate the machine through the field.

2. Throttle: The throttle control regulates the engine speed of the combine harvester. It allows the operator to adjust the engine RPM (revolutions per minute) based on the harvesting conditions, such as crop density and terrain.

3. Gear Shift: Depending on the model and design of the combine harvester, there may be a gear shift lever or buttons to select

different gears for forward or reverse movement. This control enables the operator to adjust the speed of the harvester while in motion.

4. Header Height Control: This control adjusts the height of the header, which is the front part of the harvester that cuts and gathers the crop. It allows the operator to optimize cutting height according to the crop conditions and terrain to prevent crop damage and maximize harvesting efficiency.

5. Header Tilt Control: Some combine harvesters are equipped with a header tilt control that tilts the cutting platform forward or backward. This control helps to adapt to uneven terrain and ensures proper cutting angle for efficient harvesting.

6. Threshing and Separating Controls: These controls adjust the settings of the threshing and separating mechanisms inside the combine harvester. They regulate the speed and intensity of grain separation from the crop straw, optimizing grain retention and minimizing losses.

7. Chaff Spreader and Straw Chopper Controls: In modern combine harvesters, there may be controls for adjusting the operation of the chaff spreader and straw chopper. These controls manage the distribution of chaff and straw residue behind the harvester, reducing the risk of fire and improving field management.

8. Monitor Display: Many combine harvesters are equipped with a monitor display that provides real-time information about machine performance, crop yield, grain moisture levels, and other important parameters. Operators can use this display to monitor and optimize harvesting operations.

9. Emergency Stop Button: A prominent emergency stop button is usually installed within easy reach of the operator in case of emergencies or sudden hazards. Pressing this button immediately halts all machine functions to prevent accidents or injuries.

10. Lights and Wipers: Controls for cabin lights, exterior lights, and windshield wipers ensure visibility and safety during night operations or inclement weather conditions.

These are some of the common controls found in the cabin of a combine harvester, but specific models may have additional features or variations based on manufacturer and technology advancements. Proper training and familiarity with these controls are essential for safe and efficient operation of the combine harvester.

Figure 63: Combine harvester harvesting joystick control. Ben Sutherland, CC BY-SA 2.0, via Flikr.

As an example of cabin features and controls, Figure 64outlines a typical layout.

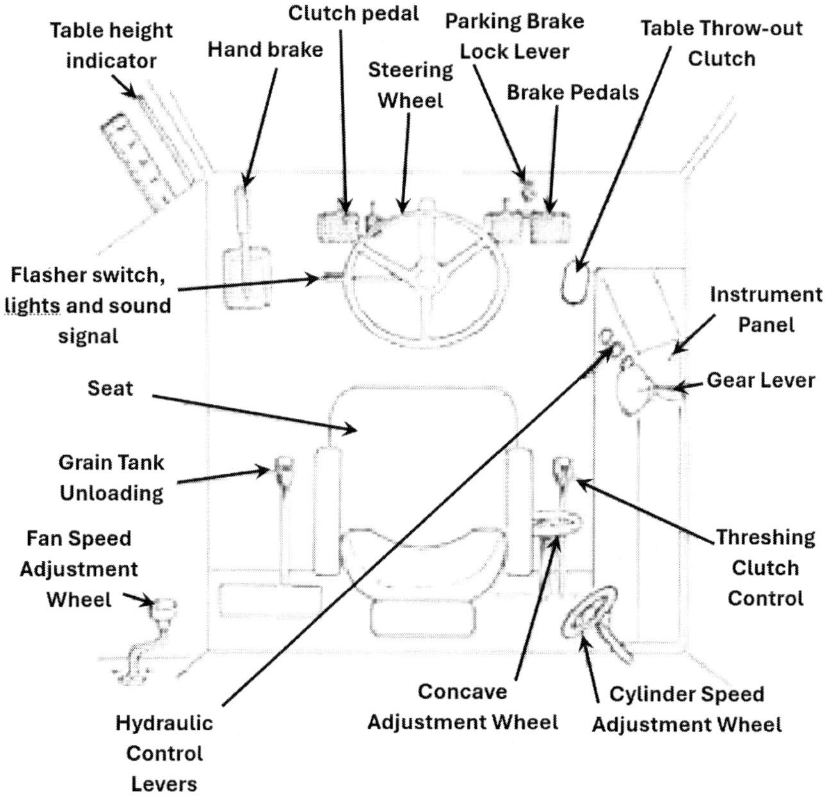

Figure 64: Example of combine harvester cabin controls.

Returning to the wheat harvesting process example, harvesting involves:

1: Adjust the Combine Header

- Ensure the combine header is appropriately adjusted relative to the height of the wheat for efficient cutting.
- Aim to leave 8 to 12 inches of wheat stubble to retain soil moisture.
- Continuously adjust the header height as the wheat height varies in the field to optimize cutting performance.

- If excessive straw is being taken in, slightly raise the header to achieve a better balance.

2: Adjust Reel Speed Relative to Ground Speed

- Fine-tune the reel speed in relation to ground speed to prevent wheat loss during harvesting.
- Avoid excessive speed, which may result in wheat knockdown or poor cutting, or overly slow speeds that could cause wheat to fall or not enter the combine correctly.
- Regularly monitor for grain loss behind the combine, indicating an imbalance between ground speed and reel speed.
- Consult the combine's manual for recommended settings to minimize grain loss.

3: Set Rotor or Cylinder Speed to Optimal Level

- Adjust rotor or cylinder speed to the minimum required for effective threshing, minimizing seed damage.
- Be prepared to modify speed settings as wheat crops change, optimizing the grain-separation process.
- Lower speeds reduce wheat damage, requiring some trial and error to find the optimal speed in the field.

4: Adjust Concave Clearance for Effective Separation

- Set the concave at the widest possible setting to aid in grain separation.
- Ensure the concave clearance is adjusted to prevent grain cracking, adjusting as needed based on crop conditions.
- The combine will automatically separate and transfer grain to the grain tank.

5: Fine-tune the Cleaning Shoe

- Adjust the cleaning shoe, comprising the chaffer and cleaning sieve, to achieve an optimal width.

- Avoid setting it too narrow or too wide, consulting the owner's manual for manufacturer-recommended settings.

- Higher grain volumes necessitate wider settings on the sieve.

6: Set the Fan

- Adjust the fan speed to ensure proper grain movement through the chaffer.

- Avoid setting the fan too low, preventing wheat from reaching the back of the chaffer, or too high, which may blow light wheat out of the shoe.

- Initially set the fan speed high and adjust as necessary to balance cleaning efficiency and grain retention.

7: Remain Attentive to Field Conditions

- Stay vigilant of how the machine interacts with the wheat, prepared to make adjustments such as fan speed as needed.

- Observing significant amounts of wheat on the ground signals the necessity to adjust settings for optimal performance.

8: Unload the Grain

- When the combine is full, unload the grain into a grain cart pulled by a tractor using the combine's unloader.

- Refer to the owner's manual for specific operation instructions tailored to your combine model.

- Utilize a separate person to drive the truck, facilitating efficient grain transportation to the storage facility while harvesting continues.

Finishing up Combine Harvester Operations

Conducting shut-down procedures according to workplace procedures is a crucial aspect of combine harvester operations. Proper shut-down ensures the safety of personnel and equipment and helps prevent damage or malfunctions during storage. Operators should follow established workplace protocols to safely shut down the combine harvester, which may include turning off the engine, disengaging power to the machinery, and securing all moving parts. Additionally, operators should follow manufacturer guidelines for shutting down specific systems or components, such as hydraulic systems or cutting mechanisms. Adhering to these procedures minimizes the risk of accidents and ensures that the combine harvester is ready for storage or maintenance.

Performing routine operational servicing and minor maintenance is essential for the continued reliability and performance of the combine harvester. Operators should conduct regular inspections and servicing tasks as outlined in the operation and maintenance manual or workplace procedures. This may include checking fluid levels, lubricating moving parts, inspecting belts and chains, and cleaning filters or screens. Performing routine maintenance tasks helps identify and address minor issues before they escalate into major malfunctions, reducing downtime and costly repairs. By prioritizing preventative maintenance, operators can ensure the longevity and efficiency of the combine harvester.

Identifying and reporting malfunctions, faults, irregular performance, or damage according to workplace procedures is critical for maintaining the operational integrity of the combine harvester. Operators should remain vigilant for any signs of abnormal performance, such as unusual noises, vibrations, or decreased efficiency. If any issues are detected, operators should promptly report them to supervisors or maintenance personnel and follow established protocols for documenting and addressing the problem. Timely reporting allows for prompt intervention and corrective action, minimizing downtime and ensuring the safety of personnel and equipment.

Cleaning, storing, and securing harvest machinery and equipment is essential for preserving their condition and longevity. After completing operations, operators should thoroughly clean the combine harvester to remove dirt, debris, and crop residue from all surfaces and components. This helps prevent corrosion, rust, and contamination, ensuring that the machinery remains in optimal condition for future use. Additionally, operators should follow workplace procedures for proper storage, which may include parking the combine in a designated area, covering or sheltering it from the elements, and securing it with locks or immobilizers. Proper storage practices protect the equipment from theft, vandalism, and environmental damage, prolonging its lifespan and reducing maintenance costs.

Recording and reporting harvest activities and machinery use records is essential for tracking performance, monitoring efficiency, and complying with regulatory requirements. Operators should maintain accurate records of harvest activities, including the type and quantity of crops harvested, operating hours, maintenance tasks performed, and any issues encountered. These records provide valuable insights into equipment utilization, productivity, and maintenance needs, helping inform decision-making and optimize operations. Additionally, maintaining detailed records facilitates compliance with industry regulations, quality standards, and contractual obligations, ensuring transparency and accountability in harvest operations.

Chapter Nine

Grain Carts

A grain cart is a piece of equipment commonly used in modern agricultural practices, particularly during the harvesting process. It serves as a mobile storage unit for harvested grain, allowing farmers to efficiently transport grain from the combine harvester to a waiting truck or storage facility.

Grain carts typically consist of a large hopper or bin mounted on a wheeled chassis, often with a hydraulic system for raising and lowering the hopper. The capacity of grain carts can vary widely, ranging from a few hundred bushels to over a thousand bushels, depending on the size and model.

During harvesting, the grain cart is positioned in the field near the combine harvester. As the combine harvests the crop, it unloads the harvested grain directly into the grain cart's hopper. Once the hopper is full, the grain cart can then transport the grain to a designated location for storage or further processing.

Figure 65: Corn combine harvest with grain cart. Wikideas1, CC0, via Wikimedia Commons.

Grain carts offer several advantages in agricultural operations. They increase the efficiency of the harvesting process by reducing downtime spent waiting for trucks to transport grain from the field. Additionally, grain carts help to minimize soil compaction in the field since they distribute the weight of the harvested grain over a larger area compared to traditional grain trucks.

Grain carts play a vital role in the logistics of harvesting, acting as a crucial link between the combine and the grain truck. They are also utilized during sowing seasons to transport seeds and fertilizers. Given their usage in rugged field conditions, it's imperative that grain carts are robustly built to withstand the demanding nature of agricultural activities and terrain challenges without encountering failures or requiring frequent repairs and part replacements.

There is a growing demand for high-capacity grain carts among farmers, as investing in quality equipment is a means to minimize downtime and expedite harvesting operations, consequently enhancing the efficiency of the combine. Considering this, what specifications should one consider when selecting the most suitable grain cart to meet the specific needs of their operation?

Capacity is a crucial consideration, varying depending on factors such as farm size, combine model, capacity, and regional requirements. It's essential to opt for a grain cart that aligns with the scale of the operation and functions harmoniously with the combine. The larger the combine's tank, the greater the grain cart's capacity should be.

Auger configuration is another key aspect to evaluate. While carts with two augers offer faster unloading times, those with a single auger and pyramidal design present fewer moving parts, minimizing breakage and grain loss, particularly beneficial for high-value operations such as seed harvesting. Auger diameter also influences unloading speed, with larger diameters facilitating quicker unloading but demanding more tractor power.

Unloading speed significantly impacts operational efficiency, necessitating synchronization between the combine and the cart to prevent downtime. Axle and wheel choices, whether single or twin axle, are linked to loading capacity, farming practices, and soil conditions. Opting for twin axles or flotation tyres helps distribute weight evenly, mitigating soil compaction and preserving soil porosity.

Ensuring compatibility with available tractor size and power is paramount, as the tractor must be capable of pulling a loaded cart and activating its unloading systems through a power take-off (PTO). When purchasing a grain cart, consider factors such as equipment quality, brand reputation, durability, and potential resale value.

Lastly, contemplate the growth trajectory of the agricultural operation. Choose a grain cart that aligns with future expansion plans, accommodating increased planted areas or the acquisition of high-performance harvesters. Selecting a grain cart with scalability ensures long-term operational efficiency and effectiveness.

While some grain farmers still opt for gravity flow wagons, the prevalent practice nowadays involves bypassing direct unloading into wagons. Instead, a grain cart, also referred to as an augur cart, is commonly employed to facilitate the transfer of corn from the combine to the wagon or truck positioned at the field's end. Equipped with large flotation tyres or tracks, grain carts boast remarkable manoeuvrability, allowing them to traverse nearly any terrain within the field, even in adverse conditions such as mud. Utilizing a grain cart can elevate harvest efficiency by over 25%, as it permits the combine to maintain near-continuous harvesting operations.

The primary advantage of employing a grain cart lies in its capacity to offload the harvest from the combine's hopper into the grain cart while the combine seamlessly continues its harvesting task. The tractor operator manoeuvring the grain cart strategically aligns it alongside the combine, matching its speed precisely. Once synchronized, the combine operator initiates the unloading process by pressing a designated button. As both vehicles advance, the combine's hopper is emptied in approximately two minutes. Subsequently, the grain cart operator disengages from the combine, allowing the latter to resume harvesting independently until its hopper nears full capacity once more.

After several loads, the grain cart operator transports the accumulated harvest to the end of the field, where it is subsequently unloaded into either a semi-truck or wagons. This streamlined process minimizes downtime and maximizes productivity during the crucial harvesting period.

Grain carts come in various types, each designed to meet specific agricultural needs and preferences. Some common types of grain carts include:

1. Single-Auger Grain Carts: These grain carts feature a single auger for unloading harvested grain. They are typically more compact and lighter, making them suitable for smaller operations or fields with limited space.

2. Dual-Auger Grain Carts: Dual-auger grain carts are equipped with two augers for unloading grain. This design allows for faster unloading speeds and increased efficiency, particularly in larger-scale operations where time is of the essence.

3. High-Capacity Grain Carts: High-capacity grain carts are built to handle larger volumes of grain, making them ideal for high-yield operations or farms with extensive acreage. These carts often have larger hopper capacities and can accommodate larger combines with higher grain output.

4. Track Grain Carts: Track grain carts feature tracks instead of traditional wheels, providing better traction and reduced soil compaction, particularly in wet or muddy conditions. They are well-suited for farms with challenging terrain or those looking to minimize soil damage.

5. Flotation Tyre Grain Carts: Flotation tyre grain carts are equipped with large flotation tyres that distribute weight evenly and reduce soil compaction. These carts are versatile and can operate in various field conditions, making them popular among farmers seeking flexibility.

6. Folding Grain Carts: Folding grain carts feature hoppers or booms that can be folded or extended, allowing for easier transport and storage when not in use. This design saves space and increases manoeuvrability, particularly when navigating narrow roads or gateways.

7. Specialty Grain Carts: Some grain carts are designed for specific purposes or tailored to unique farming needs. For example, there are grain carts with built-in scales for precise measurement, grain carts with adjustable spouts for targeted grain distribution, and grain carts with hydraulic drive systems for enhanced efficiency.

Overall, the choice of grain cart type depends on factors such as farm size, harvesting capacity, field conditions, and individual preferences. By selecting the most suitable grain cart type, farmers can optimize their harvesting operations and improve overall productivity.

Figure 66: Single auger grain cart, filled by harvester and towed by tractor. David Wright / Harvesting on Saxby Carrs, CC BY-SA 2.0, via Wikimedia Commons.

A grain cart typically consists of several key components designed to facilitate the efficient and safe transport of grain during harvest operations. These components may vary slightly depending on the specific model and manufacturer, but generally include:

1. Hopper: The hopper is the main storage compartment of the grain cart where harvested grain is deposited. It is typically located above the wheels and auger assembly and is designed to hold a large volume of grain.

2. Auger Assembly: The auger assembly consists of an auger or conveyor system that extends from the bottom of the hopper to facilitate the unloading of grain from the cart. The auger is powered by a hydraulic motor or power take-off (PTO) shaft and is used to transfer grain from the cart to a truck or storage facility.

3. Frame: The frame provides the structural support for the grain cart and is usually made of heavy-duty steel or other durable materials. It is designed to withstand the weight of the grain as well as the stress and strain of field operations.

4. Wheels or Tracks: Grain carts are typically equipped with either wheels or tracks to facilitate movement across the field. Wheels may be standard or flotation tyres, while tracks provide better traction and flotation in wet or muddy conditions.

5. Hitch: The hitch is the connection point between the grain cart and the towing tractor. It allows the cart to be securely attached to the tractor and provides a means for transferring power from the tractor to the cart's auger assembly.

6. Hydraulic System: Grain carts are equipped with a hydraulic system that powers the operation of the auger assembly, as well as other functions such as folding and unfolding the auger, raising and lowering the hopper, and controlling the cart's steering and braking systems.

7. Control Console: Many modern grain carts feature a control console located near the hitch or on the side of the cart. This

console allows the operator to monitor and control various functions of the cart, such as auger speed, hopper height, and hydraulic operation.

8. Safety Features: Grain carts may be equipped with various safety features to protect operators and prevent accidents. These may include safety shields and guards, emergency stop buttons, safety chains, and reflective markings or lighting for visibility.

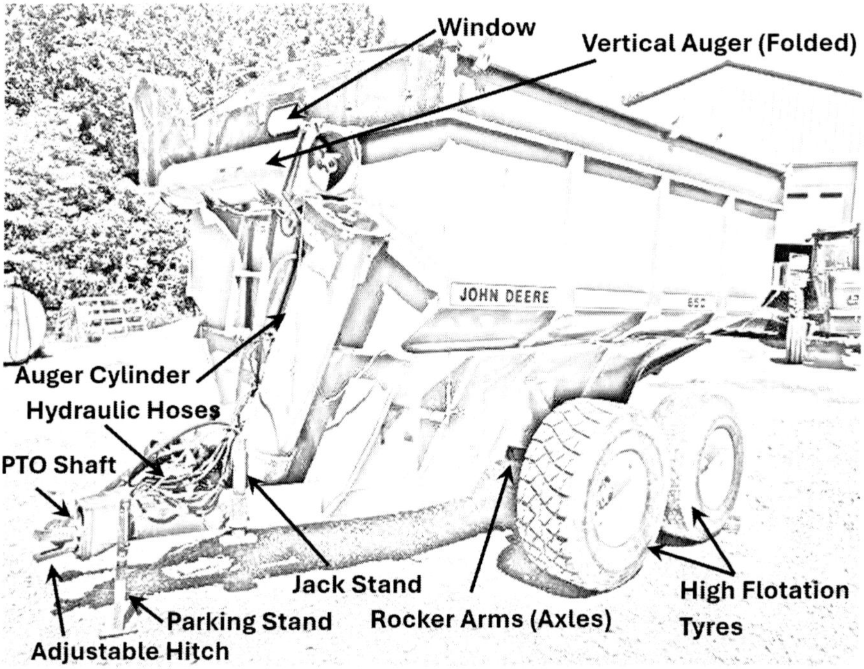

Figure 67: Grain Cart components.

Planning and Preparing for Grain Cart Operations

Planning for and preparing for grain cart operations is paramount to ensuring efficiency, safety, and the overall success of harvest operations. Here's a comprehensive guide on how to plan for and prepare for grain cart operations:

Assess Field Conditions: Begin by thoroughly assessing the field conditions before initiating grain cart operations. Consider variables such as field size, terrain features, soil moisture levels, and potential obstacles or hazards. This assessment will facilitate the identification of optimal routes for grain cart movement and highlight any areas requiring special attention or caution.

To assess field conditions before initiating grain cart operations, follow these steps:

1. Field Size: Determine the size of the field where grain cart operations will take place. Measure the dimensions of the field to understand its overall area and layout.

2. Terrain Features: Take note of any terrain features present in the field, such as slopes, hills, valleys, or uneven ground. Assess how these features may impact grain cart movement and navigation.

3. Soil Moisture Levels: Evaluate the moisture content of the soil throughout the field. Use soil moisture meters or visually inspect the soil to determine its moisture level. Consider how soil moisture may affect traction and manoeuvrability for the grain cart.

4. Potential Obstacles or Hazards: Identify any potential obstacles or hazards within the field that could impede grain cart operations. This may include rocks, tree stumps, ditches, or other obstructions. Assess the severity of these obstacles and plan accordingly to navigate around them safely.

5. Optimal Routes: Based on the assessment of field size, terrain features, soil moisture levels, and potential obstacles, identify optimal routes for grain cart movement. Determine the most efficient paths that minimize travel time and maximize productivity. Consider factors such as straight paths, level terrain, and minimal obstacles when planning routes.

6. Areas Requiring Attention: Highlight any areas within the field that may require special attention or caution during grain cart operations. This could include areas with particularly steep slopes, soft soil conditions, or dense vegetation. Develop strategies to safely navigate these areas and mitigate potential risks.

By assessing field conditions before initiating grain cart operations, you can identify optimal routes for movement, mitigate potential risks, and ensure safe and efficient operation of the grain cart in the field.

To determine the optimal routes for grain cart movement, it's crucial to consider various factors such as field size, terrain features, soil moisture levels, and potential obstacles. Here's a detailed guide on how to identify and plan these optimal routes:

Firstly, assess the field size by gaining an understanding of its overall layout and dimensions. Measure the boundaries and shape of the field to visualize the area that the grain cart needs to cover.

Next, analyse the terrain features present in the field. Take note of slopes, hills, valleys, or uneven ground that could affect the movement of the grain cart. Steep slopes may require cautious navigation, whereas level terrain can facilitate smoother travel.

Evaluate the soil moisture levels across different areas of the field. Use soil moisture meters or visual inspection to determine the moisture content. Wet or muddy soil conditions can impact traction and manoeuvrability, so it's crucial to consider this when planning routes.

Identify potential obstacles or hazards within the field that could impede grain cart operations. This includes rocks, tree stumps, ditches, or other obstructions. Assess the severity of these obstacles and plan how to navigate around them safely.

Prioritize straight paths when planning routes to minimize unnecessary turns and detours. Straight paths enable efficient movement and reduce travel time between pickup and drop-off points.

Opt for routes that traverse level terrain whenever possible, as it provides better stability for the grain cart and reduces the risk of tipping or getting stuck. Avoid excessively rough or uneven terrain that could pose challenges for cart movement.

Choose routes with minimal obstacles or hazards along the way. Clear paths of rocks, debris, or other obstructions to ensure smooth and uninterrupted movement of the grain cart. Utilize existing tracks or pathways in the field to minimize the need for clearing obstacles.

Ultimately, prioritize routes that maximize productivity and efficiency during grain cart operations. Aim to minimize travel time between pickup and drop-off points while ensuring safe and smooth movement of the cart. By prioritizing productivity, harvest operations can be optimized, and overall efficiency can be maximized.

Coordinate with Harvesting Team: Establish clear communication channels with the harvesting team, including combine operators and other equipment personnel, to coordinate schedules and logistical arrangements. Discuss the planned harvesting sequence, designate pickup points, and address any specific instructions or preferences for grain cart operations.

Review Safety Procedures: Ensure that both yourself and your team are well-versed in safety procedures and protocols for grain cart operations. Provide comprehensive training on safe operating practices, including equipment usage, emergency procedures, and hazard recognition. Emphasize the importance of maintaining a safe distance from moving machinery and adhering to speed and weight restrictions.

Inspect and Prepare Equipment: Conduct a meticulous inspection of the grain cart and tractor prior to commencing operations. Check for any signs of damage, wear, or malfunction, and promptly address any identified issues. Ensure that all safety features, such as guards, shields, and emergency stop buttons, are intact and functioning correctly. Perform routine maintenance tasks, including lubrication and bolt tightening, to keep the equipment in optimal working condition.

Load Grain Cart with Supplies: Stock the grain cart with essential supplies and equipment, such as fuel, hydraulic fluid, grease, and spare parts. Equip the grain cart with any necessary tools or accessories required for field repairs or maintenance tasks. Verify the functionality of communication devices, such as two-way radios or cell phones, to ensure seamless communication during operations.

Plan Grain Cart Routes: Develop a comprehensive plan for grain cart routes and pickup points based on factors such as field layout, harvesting sequence, and logistical considerations. Identify designated staging areas for grain cart loading and unloading, as well as safe turnaround points and access routes. Take into account field boundaries, obstacles, and soil conditions when charting grain cart routes.

Monitor Weather Conditions: Stay vigilant and monitor weather forecasts and conditions leading up to and during grain cart operations. Be prepared to adapt plans and schedules in response to changing weather conditions, such as rain, wind, or storms. Take necessary precautions to minimize risks associated with adverse weather, such as avoiding operations in wet or muddy fields or seeking shelter during lightning storms.

Communicate and Coordinate: Maintain open and clear communication channels with the harvesting team throughout grain cart operations. Utilize two-way radios, hand signals, or other communication methods to relay instructions, coordinate movements, and address any issues or concerns promptly. Foster a collaborative environment where all team members are engaged and informed about the progress and status of operations.

Grain cart operation entails various hazards and risks, which, if not managed properly, can lead to accidents, injuries, or damage to equipment. Some common hazards associated with grain cart operation include:

1. Entanglement and Entrapment: The moving parts of grain carts, such as augers and belts, pose entanglement hazards to operators and bystanders. Loose clothing, jewellery, or body parts can get caught in these mechanisms, resulting in serious injuries or fatalities.

2. Falls and Slips: Working on or around grain carts, especially when loading or unloading, increases the risk of slips, trips, and falls. Wet or uneven surfaces, spills, or debris can contribute to accidents if not properly managed.

3. Struck-By Accidents: Grain carts are heavy pieces of equipment, and being struck by them or their moving parts poses a significant risk to operators and bystanders. Accidents can occur if individuals are caught in the path of a moving grain cart or are struck by swinging augers or other components.

4. Machinery Entanglement: Grain cart operators may inadvertently come into contact with other machinery, such as combines or tractors, leading to machinery entanglement accidents. Failure to maintain a safe distance or failing to communicate effectively with other operators increases the risk of such incidents.

5. Overturning: Grain carts can overturn if operated on uneven terrain or at high speeds, particularly when fully loaded. Overturning accidents pose serious risks to operators and bystanders, as well as potential damage to equipment and spilled grain.

To mitigate these hazards and risks associated with grain cart operation, several measures can be implemented:

1. Operator Training: Provide comprehensive training to operators on safe grain cart operation practices, including proper use of controls, emergency procedures, and hazard recognition. Ensure operators are aware of the risks associated with grain cart operation and understand how to mitigate them.

2. Personal Protective Equipment (PPE): Require operators to wear appropriate PPE, such as high-visibility clothing, steel-toed boots, gloves, and hearing protection, to protect against potential hazards, including entanglement, falls, and struck-by accidents.

3. Equipment Inspection and Maintenance: Conduct regular inspections of grain carts to identify and address any mechanical issues or defects promptly. Ensure all safety features, such as guards, shields, and emergency stop buttons, are in place and functioning correctly. Perform routine maintenance tasks, such as lubrication and tightening of bolts, to keep equipment in optimal condition.

4. Safe Operating Procedures: Develop and enforce safe operating procedures for grain cart operation, including guidelines for loading and unloading, manoeuvring in the field, and communicating with other operators. Emphasize the importance of maintaining a safe distance from moving machinery and adhering to speed limits and weight restrictions.

5. Communication and Coordination: Establish clear communication protocols among operators, especially when working in teams or alongside other equipment. Use two-way radios or hand signals to communicate effectively and coordinate movements to minimize the risk of accidents or collisions.

6. Terrain Assessment and Planning: Conduct a thorough assessment of field terrain and conditions before grain cart operation. Identify potential hazards, such as steep slopes, ditches, or soft ground, and take appropriate precautions to mitigate risks, such as avoiding high-risk areas or using alternative routes.

7. Emergency Preparedness: Ensure operators are trained in emergency response procedures, including how to shut down equipment safely in case of emergencies and how to administer first aid if injuries occur. Keep emergency contact information readily available and establish protocols for reporting accidents or incidents.

Preparing the grain cart in advance can minimize downtime and enhance the safety and efficiency of the harvest. Utilize the following checklist as a guide while inspecting your grain cart before the season begins.

Visual Inspection: Conduct a visual assessment of your grain cart, checking for any broken or damaged parts, cracked welds, or debris remaining from the previous season. Replace or repair any damaged components and clear debris from the cart.

Wheel Nuts: Inspect all wheel nuts to ensure they are tightened to the correct torque setting. Torque ¾" nuts to 400 ft-lbs and 22mm nuts to 640 ft-lbs.

Hitch: Ensure that the tractor pin fits properly into your grain cart hitch. For a Standard Hitch, ensure that the tractor pin is at least ¼" smaller in diameter than the pin hole in the grain cart hitch to allow sufficient free movement on uneven terrain.

Power Take-Off (PTO): Check the telescoping PTO for adequate extension. With the tractor hitched to the cart on level ground, the PTO should extend approximately 10 inches from its collapsed length. If equipped with a Comer PTO, verify correct adjustment of the PTO slip clutch. For grain carts equipped with an 18", 20", or 22" diameter auger, adjust the spring length as per specifications.

Hydraulic Cylinders & Hoses: Operate all hydraulic cylinders and inspect fittings and seals for leaks. Use cardboard or wood to detect hydraulic leaks.

Lubrication: Consult the operator's manual to verify the gearbox oil level. Exercise horizontal and vertical cleanout doors and lubricate hinge points if necessary.

Lighting & Safety Features: Connect the grain cart wire harness to the tractor receptacle and ensure that all lighting functions properly. Activate the auger spout light using the designated button or switch. Verify that safety chains are correctly installed and guards/shields are in place.

Tyres: Check that all tyres are inflated to the specified air pressure indicated in the operator's manual.

Tracks: Inspect the belt for any defects. Check the tension and alignment of belts, referring to the track's operator's manual for adjustment procedures. Follow the operator's manual for lubricating grease points and oil bath hubs.

Operating a Grain Cart

Effectively operating the tractor and grain cart demands skill and entails a rapid-paced responsibility. The cart operator remains constantly engaged in various tasks, whether it's fetching grain, shuttling to and from the truck, or unloading the harvested grain. Anticipation and forward-thinking are key attributes required for this role, as the operator must foresee and plan their movements accordingly, aiming to synchronize with the combine's operations seamlessly.

Figure 68: Harvester filling grain cart. Michael Trolove / Harvest Action, CC BY-SA 2.0, via Wikimedia Commons.

While traversing the field, the operator's approach should prioritize reaching the location where the combine will be, rather than its current position. The ultimate objective is to ensure the continuous operation of the combine, minimizing any interruptions to the harvesting process. This strategic approach allows for optimal efficiency and productivity throughout the harvesting endeavour.

When navigating curves or hills, ensure to maintain a low speed and ensure that at least 20% of the tractor's weight is distributed on the front wheels to ensure safe steering. When encountering rough or uneven terrain, reduce speed accordingly. Exercise caution whenever the unloading auger is extended to avoid proximity to overhead power lines, as electrocution can happen even without direct contact.

Positioning and coordination of the grain cart in the field is a critical aspect of efficient harvest operations. This includes:

Field Assessment: Begin by assessing the layout of the field and identifying key features such as the location of the combine, grain trucks, and access routes. Take note of any obstacles or hazards that may affect the movement of the grain cart.

Combine Location: Determine the current location of the combine and anticipate its path as it moves through the field during harvesting. Position the grain cart strategically to intercept the combine at optimal points along its route.

Grain Truck Access: Consider the location of the grain trucks and access points for unloading harvested grain. Position the grain cart in proximity to these locations to minimize travel distance and streamline the loading process.

Optimal Pickup and Drop-Off Points: Coordinate with the harvesting team, including combine operators and grain truck drivers, to establish optimal pickup and drop-off points for the harvested grain. Determine designated areas within the field where the grain cart will be stationed for loading and unloading operations.

Communication with Team: Maintain clear communication with the harvesting team to ensure alignment of objectives and efficient coordination of activities. Use two-way radios, hand signals, or other communication methods to relay instructions, coordinate movements, and address any logistical challenges.

Flexibility and Adaptability: Remain flexible and adaptable in response to changing field conditions or unforeseen circumstances. Be prepared to adjust the positioning of the grain cart as needed to optimize workflow and accommodate evolving harvesting requirements.

Safety Considerations: Prioritize safety throughout the positioning and coordination process. Ensure that the grain cart is positioned away from hazards such as uneven terrain, overhead power lines, or other obstacles. Maintain a safe distance from moving machinery and adhere to established safety protocols at all times.

Efficiency and Productivity: Strive to maximize efficiency and productivity by positioning the grain cart strategically to minimize downtime and streamline operations. Aim to optimize the flow of harvested grain from the combine to the grain trucks or storage facility while minimizing unnecessary travel and delays.

By carefully considering factors such as combine location, grain truck access, and communication with the harvesting team, you can effectively position and coordinate the grain cart in the field to support a smooth and efficient harvest operation.

Monitoring grain flow during harvest operations is crucial for ensuring efficient loading and unloading processes entails:

Closely Monitor Grain Flow: Keep a close eye on the grain flow from the combine to the grain cart. Pay attention to the rate at which grain is being harvested and transferred to the cart. This allows you to anticipate when the grain cart will need to be moved to accommodate the incoming grain.

Be Prepared to Adjust Positioning: Stay proactive and ready to adjust the positioning of the grain cart as necessary. If the grain flow from the combine increases or decreases, be prepared to move the grain cart closer or farther away to maintain optimal loading conditions. This helps prevent delays and ensures a steady flow of grain.

Communicate with Combine Operator: Maintain open communication with the combine operator to coordinate movements effectively. Inform the combine operator of any adjustments you make to the grain cart's positioning and work together to synchronize movements. This collaboration helps optimize productivity and minimize downtime.

Optimize Loading and Unloading Operations: Aim to optimize loading and unloading operations by keeping the grain flow steady and consistent. Coordinate with the combine operator to ensure that the grain cart is positioned in the right location at the right time to receive harvested grain efficiently. This helps maximize throughput and overall harvest efficiency.

By closely monitoring grain flow, being prepared to adjust positioning as needed, communicating effectively with the combine operator, and optimizing loading and unloading operations, you can ensure smooth and efficient harvest operations with the grain cart.

Transport to Designated Unloading Point: Once the grain cart reaches full capacity, carefully transport it to the designated unloading point, which may be a nearby grain truck or a storage facility. Follow designated routes and exercise caution to ensure safe transportation.

Figure 69: Grain cart positioned for unloading. Cd design85, CC BY-SA 3.0, via Wikimedia Commons.

Prepare for Unloading: Before initiating the unloading process, prepare the grain cart for unloading by ensuring that it is parked on stable ground and properly aligned with the unloading point. Engage any safety mechanisms or locking mechanisms to secure the cart in place during unloading.

Figure 70: Positioning for emptying the grain cart. Judd McCullum, CC BY-SA 2.0, via Wikimedia Commons.

Emptying a grain cart involves transferring the harvested grain from the cart's hopper to a storage facility or another vehicle, typically a truck or grain wagon. Here's a step-by-step guide on how to empty a grain cart:

1. Position the Grain Cart: Drive the grain cart to the designated unloading area, ensuring that it is positioned close enough to the receiving vehicle or storage facility for efficient unloading. Take into account factors such as terrain, ground stability, and accessibility when selecting the unloading spot.

2. Prepare the Receiving Vehicle: If unloading into a truck or grain wagon, ensure that it is properly positioned and stationary. Lower any unloading augers or gates on the receiving vehicle to facilitate the transfer of grain from the grain cart.

3. Activate Hydraulic System: Engage the hydraulic system on the

grain cart to power the unloading process. Depending on the specific configuration of the grain cart, this may involve activating a hydraulic motor or power take-off (PTO) shaft that drives the auger assembly.

4. Extend the Unloading Auger: Extend the unloading auger from the bottom of the grain cart's hopper down to the receiving vehicle or storage facility. Ensure that the auger is properly aligned and positioned to direct the flow of grain accurately.

5. Control Unloading Speed: Adjust the speed of the unloading auger to control the rate at which grain is transferred from the grain cart to the receiving vehicle. Monitor the flow of grain to prevent overfilling and spillage.

6. Monitor Grain Level: Keep an eye on the grain level in both the grain cart's hopper and the receiving vehicle or storage facility. Avoid overfilling the receiving vehicle or creating a pile of grain that exceeds its capacity.

7. Complete the Unloading Process: Continue unloading grain from the grain cart until the hopper is empty or the desired amount of grain has been transferred. Once unloading is complete, retract the unloading auger and deactivate the hydraulic system.

8. Secure the Grain Cart: After emptying the grain cart, retract any augers or gates and ensure that all components are properly secured for transport. Double-check that the grain cart is securely hitched to the towing vehicle before moving it from the unloading area.

9. Clean Up: Take the opportunity to clean any spilled grain or debris from the grain cart and surrounding area. This helps

maintain equipment cleanliness and prevents potential hazards during future operations.

Figure 71: Emptying a grain cart. IowaGoatWhisperer, CC BY-SA 4.0 , via Wikimedia Commons.

Completing Grain Cart Operations

Completing grain cart operations involves several steps to ensure that the equipment is properly packed up and stored after use. Here's a detailed guide on how to complete grain cart operations, pack up, and store the grain cart:

Firstly, ensure that the grain cart has been completely emptied of all harvested grain before proceeding with the packing-up process. Utilize any remaining grain in the cart to fill up the receiving vehicle or storage facility entirely. Once the unloading process is complete, retract the unloading auger back into its storage position and securely lock it to prevent any accidental extension during transport.

Next, turn off the hydraulic system powering the grain cart's functions, including any hydraulic motors, power take-off (PTO) shafts, or other hydraulic components used during the unloading process. Inspect the grain cart for any remaining grain residue or debris, both inside the hopper and on external surfaces. Use a broom, brush, or compressed air to remove any accumulated residue and ensure that the grain cart is clean and free of obstructions.

If the grain cart is equipped with fold-down auger extensions for increased capacity, fold them down to their storage position and secure them in place using any locking mechanisms or fasteners provided. Check that all loose components, such as safety chains, lights, or access panels, are properly secured and stowed away. Tighten any bolts or fasteners that may have loosened during operation to prevent them from becoming hazards during transport.

Conduct a final visual inspection of the grain cart to ensure that everything is in order and ready for transport or storage. Look for any signs of damage, wear, or mechanical issues that may require attention before the next use. If transporting the grain cart on public roads, ensure that it is equipped with transport lights and reflectors as required by local regulations. Attach these lights and reflectors to the grain cart's frame or hitch according to manufacturer instructions.

Figure 72: Grain cart storage. Kate Nicol / Broughton Carrs, CC BY-SA 2.0, via Wikimedia Commons.cription automatically generated

Attach the grain cart to the towing vehicle using the appropriate hitching mechanism and ensure that the hitch is securely locked in place. Transport the grain cart to its designated storage location, choosing a clean, dry, and sheltered area to protect it from the elements and minimize the risk of damage or corrosion. Before storing the grain cart, perform any necessary routine maintenance tasks, such as lubricating moving parts, checking fluid levels, or replacing worn components.

Keep all manuals, documentation, and maintenance records for the grain cart in a safe and easily accessible location to ensure that important information is readily available for reference or troubleshooting as needed. By following these steps, you can safely and effectively complete grain cart operations, pack up, and store the grain cart until it is needed for the next harvest season.

Chapter Ten

Balers and Bale Wrappers

An agricultural baler is a specific type of baler used primarily in farming operations to compress and bundle agricultural materials such as hay, straw, and other forage crops into compact and manageable bales. These bales are then typically used for livestock feed, bedding, or other agricultural purposes.

A baler apparatus is a compacting mechanism employing a ram to compress materials into rectangular bales for shipping, storage, processing, or recycling. Engineered to handle both similar and dissimilar waste, they transform them into manageable configurations, maximizing space and optimizing resources.

Figure 73: New Holland BR7060 hay baler. Acroterion, CC BY-SA 4.0, via Wikimedia Commons.

Agricultural balers come in various designs, including round balers and square balers.

1. Round Balers: Round balers create cylindrical bales by rolling the material into a round shape. They are commonly used for baling hay and straw. Round balers are efficient and can handle a high volume of material, making them popular for larger-scale farming operations.

2. Square Balers: Square balers compress the material into rectangular or square-shaped bales. These balers are typically used for baling hay, straw, and sometimes even corn stalks. Square bales are often preferred for their uniform shape, stackability, and ease of transport and storage.

Agricultural balers play a crucial role in modern farming operations by helping farmers efficiently harvest, handle, and store their crops for feeding livestock, selling, or other uses. They increase efficiency, reduce manual labour, and facilitate the management of agricultural materials.

Figure 74: Square baler and accumulator. Glendon Kuhns, CC0, via Wikimedia Commons.

Agricultural produce and waste are baled to retain their value and render them suitable for proper disposal or utilization as raw materials. Animal feed, such as hay, grass, and straw, serves as fodder for domesticated livestock, aiming to fulfill their nutritional requirements with high energy and protein content. However, animal fodder, being voluminous and prone to decomposition, necessitates measures such as drying and shaping. Additionally, as fodder decays, heat is generated, posing a risk of spontaneous combustion. Baling aids in regulating the moisture levels of the fodder, as insufficient moisture leads to dryness and loss of nutritional quality, while excess moisture can lead to spoilage and subsequent combustion. Typically, baled animal fodder is encased in plastic sheets or covers to aid in moisture retention.

Round Hay Balers, also known as cylindrical hay balers, specialize in producing round bales. The weight of these bales varies depending on their size, typically 4ft or 5ft, and whether they are compacted as 'hardcore' or 'softcore' bales. This distinction signifies the degree of compression, significantly affecting the bale's weight.

Figure 75: Round hay bales produced by round baler. Mtaylor848, CC BY-SA 4.0, via Wikimedia Commons.

In operation, the round baler gathers hay from the ground and channels it into the bale chamber. The hay is then wrapped around itself using six to eight elongated rubber belts, each approximately 7 inches wide. As more hay is collected and fed into the baling mechanism, the bale takes on a spherical shape, filling the bale chamber. The pressure exerted by the hay on the belts is monitored by a hydraulic system. Upon reaching a predetermined tension level, a signal is sent to the tractor operator, who halts the tractor. The bale is then wrapped with baler twine or protective sheeting. Following wrapping, the pressure on the belts is released, and the rear section of the baler is opened via hydraulic cylinders, allowing the bale to roll out onto the ground. The tractor advances, closing the rear section, and baling resumes.

Rectangular Hay Balers, commonly known as square balers, produce bales with rectangular dimensions. While these bales can vary in length, typical sizes include 8' for straw, 7' for hay, and around 4' for haylage/silage bales wrapped in protective film. Due to their weight and size, a tractor or loader is necessary for moving and stacking them. However, similar to smaller bales, they are compacted into 'biscuits' or 'slices,' enabling easy transportation for feeding animals.

Figure 76: New Holland D1010 hay baler. David Hawgood / New Holland D1010 hay baler, CC BY-SA 2.0, via Wikimedia Commons.

In the process, hay enters the baler through a pickup mechanism, with teeth gently raking it from the ground to prevent debris from entering. A compression bar located just behind the pickup secures the hay in place, allowing the auger to transfer it into the bale chamber.

The bale is then compressed and cut to achieve the desired size and shape. Once the appropriate length and size are attained, the bale is wrapped with two lengths of baler twine or wire and securely fastened. Subsequently, the bale is released onto the bale chute, falling to the ground, or propelled onto a hay rack by kickers or bale ejectors on certain models.

The essential functional elements that constitute a rectangular baler for forming bales are the pickup and elevating unit, feed conveyor, feeder, compression chamber, and tying mechanism. Each of these components will be elaborated upon individually.

Pickup and Elevating Unit: This unit retrieves straw from the windrow and raises it to a level where a feed conveyor can transport it to the bale chamber. The pickup comprises spring-loaded tines that delicately lift the windrow and guide it along the stripper bars. At the top of the pickup, the tines retract downward to prepare for lifting a new section of the windrow. In some instances, the pickup also functions as the elevator, while older models might feature a separate elevator. Balers designed for gathering straw from windrows formed by mower conditioners may feature pickup tines spaced closer together.

Feed Conveyor: This mechanism transfers straw to the side of the bale chamber. Large-diameter augers are commonly used for feed conveyors. However, in certain cases, several extenders from the feeder serve to convey the material to the side of the bale chamber. On older models, rubber belts were employed for this purpose.

Feeder: The feeder accepts straw from the end of the feed conveyor and deposits it into the bale chamber. Typically, the feeder enters the bale chamber when the plunger is in the forward position, necessitating precise timing with respect to the plunger.

Compression Chamber: Straw undergoes compression within the bale chamber, facilitated by the plunger driven by a sizable crankshaft and connecting rod. Each new batch of straw is compressed against the preceding charges, gradually forming the bale as it moves through the chamber under the plunger's force.

Tying Mechanism: The tying mechanism comprises knotters, needles, and a metering wheel. These components are responsible for tying each bale when it reaches the desired length. The fingers of the metering wheel partially extend into the bale chamber, causing it to rotate slowly as the bales traverse the chamber. Rotation of the metering wheel activates the knotter clutch, engaging power to the needles and knotters to tie the bale. The relationship between the metering wheel and the knotter clutch can be adjusted to alter the bale length, typically by shifting a collar on the trip arm at the metering wheel.

Rectangular Pickup Baler: This type of baler forms bales from straw windrows left by combines. Principally, the machine comprises a pickup reel, conveying and feeding system, compression chamber, bale density adjuster, bale length controller, needle and tying mechanism, crank linkage mechanism, and power transmission and hauling system. It is powered via the power take-off of the hauling tractor. The pickup reel with spring teeth lifts the straw windrow from the ground and continuously transfers it to the conveying and feeding mechanism as the baler advances along the windrow.

Figure 77: Rectangular baler components. Back image - Acroterion, CC BY-SA 4.0, via Wikimedia Commons.

Round balers designed for straw collection are typically PTO operated and are towed directly behind the tractor. It is preferable to utilize a tractor with a wide front axle to straddle the windrow effectively. To ensure round bales of uniform diameter, large and wide windrows are favoured. In instances where smaller windrows are harvested, operators often employ a weaving motion while the bale is being formed to prevent tapered bales.

These round balers can be categorized based on their working unit into long-belt type, short-belt type, chain type, and roller type. Additionally, they can be classified by their working principle as inside winding type and outside winding type. Long-belt and chain types are inside winding, while short-belt and roller types are outside winding. Furthermore, balers can be categorized based on chamber size adjustment as expandable chamber round pickup balers, ground roll balers, and fixed volume round pickup balers.

1. Expandable Chamber Round Pickup Baler: This is the most

prevalent type of round baler. It employs a conventional tooth pickup to collect the straw windrow and transfers it into the bale chamber using rollers and belts. The straw is compressed using belts and rollers or apron chains. As straw is fed into the bale chamber, it expands, resulting in a bale of relatively uniform density. Once the bale reaches the desired size, twine is fed into the chamber and wrapped around the bale as it rotates. After approximately one revolution, forward movement halts, and twine continues to be fed while the bale makes 6 to 10 more revolutions. Upon completion of wrapping, the tailgate is raised, and the bale is ejected. The process repeats with the lowering of the tailgate to begin a new bale. These balers typically produce high-density bales weighing between 0.5 to 1.0 tonne. They have a capacity ranging from 2 to 12 t/h, with power requirements varying from 10 to 25 kW, although a tractor with at least 48 kW is recommended for optimal utilization.

2. Ground Roll Baler: Ground roll balers utilize the least tractor horsepower among the three types. A pickup is employed to roll the straw forward along the ground, while belts, grids, or cables are used to form round bales. Bales formed by ground roll balers are generally lighter and less dense compared to those produced by other types, making them susceptible to losses during handling, transportation, and storage. Capacity with these balers largely depends on operator experience and can range from 1 to 5 t/h. Power requirements typically do not exceed 20 kW, although a tractor with at least 34 kW is recommended for optimal use.

3. Fixed Volume Round Pickup Baler: In this type of baler, the pickup lifts the straw into the bale chamber. However, the chamber has a fixed volume, and the bale does not take shape

until the chamber is nearly full. Consequently, bales produced by this machine have a lower-density core compared to those made by expandable chamber balers. Although this type of baler boasts a capacity similar to expandable chamber types, it tends to have higher power requirements, reaching up to 45 kW. Overall capacity for this baler aligns with that of the Expandable Chamber Round Pickup Baler.

Inside Winding Round Pickup Baler: This type of baler consists of a pick-up reel, conveying and feeding mechanism, wrapping and pressing mechanism, rear door for unloading, transmission mechanism, and hydraulic operating mechanism. The windrow is lifted by a pickup reel and pressed into a flat layer by double smooth rollers before being conveyed to the baler chamber. The straw moves upward by friction to a certain height and then rolls down to form the bale's core. The bale continues to roll, increasing in diameter until it reaches the desired size, at which point the binding mechanism is activated, and twine is wrapped around the bale's circumference. The bale is then discharged to the ground. Bales formed by inside wrapping balers have higher density and better shape retention during storage, although the machine's structure is more complex compared to others.

A bale wrapper, on the other hand, is a piece of agricultural equipment used to wrap bales of forage, such as hay or silage, with plastic film. The process of wrapping bales with plastic film is known as bale wrapping. Bale wrappers typically consist of a mechanism for holding and rotating the bale, as well as a system for dispensing and applying plastic film around the bale.

The primary purpose of bale wrapping is to preserve the quality and nutritional value of forage by creating an airtight seal around the bale. This helps to prevent spoilage, mold growth, and nutrient loss during storage. Bale wrappers are commonly used in livestock farming operations where high-quality forage is essential for feeding animals year-round.

There are different types of bale wrappers available, including stationary wrappers that require the bale to be brought to the wrapper, and trailed wrappers that can be towed behind a tractor to wrap bales directly in the field. Some bale wrappers also have features such as automatic film cutting and application, adjustable wrapping tension, and bale handling capabilities for improved efficiency and ease of use.

Figure 78: Kuhn Intelliwrap SW 404 large square bale wrapper. Peter Facey, CC BY-SA 2.0, via Wikimedia Commons.

Some units integrate a bale wrapper with a baler to create a single, unified piece of equipment known as a bale wrapper baler combination or integrated bale wrapper baler. This integration allows for the baling and wrapping processes to be performed sequentially without the need for separate equipment or additional handling of the bales.

In an integrated system, once the baling process is complete, the baler automatically transfers the bale to the wrapping mechanism, where it is wrapped with plastic film before being ejected. This streamlines the workflow, reduces labour requirements, and can result in increased efficiency, especially for operations that produce large volumes of bales.

Figure 79: McHale integrated baler wrapper Fusion pulled by a New Holland tractor. Lydur, CC BY 2.0, via Wikimedia Commons.

Integrated bale wrapper baler systems are commonly used in commercial agriculture and livestock farming operations where efficiency and productivity are key considerations. However, it's important to note that not all balers are designed to be integrated with bale wrappers, so compatibility should be verified before attempting to combine the two pieces of equipment.

Bales are wrapped primarily to preserve the quality and nutritional value of forage, such as hay or silage, during storage and transportation. Wrapping bales with plastic film creates an airtight seal around the forage, which helps to prevent spoilage and maintain the freshness of the forage by sealing out oxygen, which can promote mold growth and decomposition.

The airtight seal created by wrapping bales also helps to preserve the nutritional content of the forage, including vitamins, minerals, and protein. This ensures that the forage retains its quality and is suitable for feeding to livestock, thus preserving the nutritional value of the feed.

In addition to preserving quality and nutritional value, wrapping bales with plastic film reduces dry matter loss caused by exposure to air, sunlight, and moisture. By minimizing waste, wrapping bales helps to maximize the yield and value of the forage, making it a more cost-effective option for farmers.

Wrapped bales can be stored outdoors without the need for additional storage structures, such as barns or sheds. This improves storage efficiency and flexibility, allowing forage to be stored closer to where it will be used, thus reducing the need for long-distance transportation and storage costs.

Furthermore, wrapping bales with plastic film helps to compact the forage and hold it together, making it easier to handle and transport. This is particularly important for large-scale farming operations where efficiency and productivity are key considerations, as it streamlines the handling and transportation process, saving time and labor.

Overall, wrapping bales with plastic film is a widely adopted practice in agriculture, particularly in livestock farming, where high-quality forage is essential for animal nutrition and health. It offers numerous benefits, including preservation of quality and nutritional value, reduction of waste, improved storage efficiency, and facilitation of handling and transport.

A bale wrapper operates by enclosing bales of forage, such as hay or silage, with plastic film to establish an airtight seal around the bale. The procedure typically comprises several key stages.

Initially, the bale wrapper is positioned adjacent to the bale, which has been previously formed by a baler. The bale is then loaded onto the wrapper's wrapping platform or table, positioning it for the commencement of the wrapping process.

Figure 80: Tanco Round Bale Wrapping Machine Auto Wrap 1510. Harald Bischoff, CC BY 3.0, via Wikimedia Commons.

Once the bale is securely in place on the wrapper, the wrapping process is initiated. This initiation can be performed either manually by the operator or automatically, depending on the type of wrapper being utilized.

The wrapper dispenses plastic film from a roll or pre-cut sheets. This film, often constructed from stretchable material such as polyethylene, possesses the capability to be stretched and tightly wound around the bale.

As the bale rotates on the wrapping platform, the plastic film is applied to its surface. Rollers or other mechanisms guide the film around the bale, ensuring uniform coverage and a secure seal.

The tension and overlap of the plastic film can be adjusted to attain the desired level of wrapping. This adjustment is crucial for ensuring comprehensive coverage and airtight sealing, thereby preventing spoilage and maintaining quality.

Upon completion of the wrapping process, the bale is fully encased in plastic film. The wrapper may feature mechanisms to automatically cut and secure the film, or these tasks may be performed manually by the operator.

Subsequently, the wrapped bale is ejected from the wrapper's wrapping platform or table. It is then prepared for storage, transport, or immediate use, depending on the requirements of the operation.

Overall, a bale wrapper functions by tightly enveloping bales of forage with plastic film to establish an airtight seal, thus safeguarding the quality and nutritional value of the forage during storage and transportation. This automated and efficient process serves as a valuable tool for farmers and livestock producers.

Planning and Preparing for Baler Operations

Baler operations, while essential for processing and managing agricultural materials, come with inherent hazards that can pose risks to operators and bystanders. Some of the main hazards associated with baler operations include:

1. Entanglement and Entrapment: One of the most significant hazards is the risk of entanglement or entrapment of operators or bystanders in moving parts of the baler, such as belts, chains, or rollers. Accidental contact with these moving components can lead to serious injuries, including amputations or crush injuries.

2. Falls and Slips: Operators may be at risk of falls or slips when accessing or working around the baler, especially when performing maintenance tasks or clearing blockages. Uneven terrain, slippery surfaces, or cluttered work areas can increase the

likelihood of accidents.

3. Falling Objects: Baler components or bales themselves may shift or fall unexpectedly during operation, posing a risk of injury to operators or bystanders standing nearby. This risk is particularly relevant when loading or unloading bales from the baler or transporting them to storage areas.

4. Electrical Hazards: Baler machinery often relies on electrical systems to operate various components, such as motors or sensors. Malfunctioning electrical systems, damaged wiring, or exposed electrical components can pose electrocution hazards to operators or maintenance personnel.

5. Mechanical Failure: Like any mechanical equipment, balers are susceptible to mechanical failures, such as broken components, hydraulic leaks, or malfunctioning safety mechanisms. These failures can lead to sudden and unexpected movements or shutdowns, potentially causing injuries or damage to the equipment.

6. Fire and Combustion: The baling process can generate heat, especially when compressing organic materials like hay or straw. Under certain conditions, such as high moisture content or prolonged storage, bales can undergo spontaneous combustion, leading to fires that pose risks to both personnel and property.

7. Chemical Exposure: Some balers may use lubricants, hydraulic fluids, or other chemicals as part of their operation. Exposure to these chemicals, whether through leaks, spills, or improper handling, can pose health risks to operators, including skin irritation, respiratory issues, or chemical burns.

To mitigate these hazards and ensure safe baler operations, it is essential for operators to receive proper training on equipment use, maintenance procedures, and safety protocols. Regular inspections, maintenance checks, and adherence to safety guidelines provided by manufacturers are also crucial in preventing accidents and minimizing risks associated with baler operations.

Planning and preparing for baler operations is essential to ensure efficiency, safety, and successful completion of tasks. This includes:

Review Manufacturer Guidelines: Begin by familiarizing yourself with the manufacturer's guidelines and operating manual for the specific model of baler you will be using. Pay close attention to safety instructions, maintenance procedures, and recommended operating parameters.

Assess Operational Needs: Determine the scope and requirements of the baler operations, including the type and quantity of materials to be processed, the desired bale size, and any specific operational constraints or considerations.

Select Appropriate Equipment: Choose the appropriate baler equipment for the task based on factors such as the type of material to be baled, the desired bale size and shape, and the available resources and infrastructure.

Inspect Equipment: Conduct a thorough inspection of the baler equipment to ensure it is in good working condition. Check for signs of damage, wear, or malfunctioning components, and address any issues promptly before proceeding with operations.

Gather Necessary Materials: Collect all necessary materials and supplies needed for baler operations, including the materials to be baled, twine or wire for binding bales, lubricants or fluids for equipment maintenance, and any personal protective equipment (PPE) required for operators.

Prepare Work Area: Ensure the work area where the baler operations will take place is clear of obstacles, debris, or hazards that could interfere with equipment operation or pose risks to personnel. Make sure the ground is level and stable to support the baler and other equipment.

Implement Safety Measures: Establish appropriate safety measures to protect personnel and equipment during baler operations. This may include setting up designated work zones, restricting access to unauthorized personnel, providing training on safe equipment operation, and ensuring operators wear appropriate PPE, such as gloves, eye protection, and hearing protection.

Develop Contingency Plans: Anticipate potential challenges or issues that may arise during baler operations, such as equipment malfunctions, adverse weather conditions, or logistical constraints. Develop contingency plans and procedures to address these challenges and minimize disruptions to operations.

Communicate and Coordinate: Ensure clear communication with all personnel involved in baler operations to ensure everyone understands their roles and responsibilities. Coordinate activities and establish clear channels of communication to facilitate smooth and efficient workflow.

Monitor and Evaluate: Continuously monitor baler operations and performance to identify any issues or opportunities for improvement. Conduct regular evaluations to assess the effectiveness of operational procedures and identify areas for optimization or refinement.

Operating a Baler

Performing pre-start checks before using a baler is essential to ensure the equipment is in proper working condition, which helps prevent accidents, breakdowns, and damage to the machine. Here's a comprehensive guide on the pre-start checks that should be taken prior to using a baler:

1. Visual Inspection: Conduct a thorough visual inspection of the

baler, checking for any signs of damage, wear, or loose components. Look for leaks, cracks, or worn-out parts in critical areas such as the bale chamber, hydraulic lines, and drivetrain components.

2. Fluid Levels: Check the fluid levels of essential fluids such as hydraulic fluid, engine oil, and coolant. Ensure that all fluids are at the correct levels according to the manufacturer's specifications. Top up any fluids that are low or replace them if necessary.

3. Tyre Pressure: Inspect the tyres on the baler and ensure they are properly inflated to the recommended pressure. Check for any cuts, punctures, or damage to the tyres that could affect performance or safety.

4. Safety Devices: Test all safety devices and features on the baler, including emergency stop switches, safety guards, and warning lights. Ensure that all safety devices are functioning correctly and that they are not obstructed or damaged.

5. Hydraulic System: Check the hydraulic system for any leaks, damaged hoses, or loose fittings. Test the operation of hydraulic cylinders, valves, and controls to ensure they are functioning properly.

6. PTO Shaft: Inspect the power take-off (PTO) shaft for any signs of damage or wear. Ensure that the PTO shaft is securely attached to both the baler and the tractor and that it rotates smoothly without any binding or unusual noises.

7. Bale Chamber: Open the bale chamber and inspect the interior for any debris, obstructions, or signs of damage. Ensure that the chamber is clean and free from any foreign objects that could interfere with the baling process.

8. Electrical System: Check the electrical system, including wiring, connectors, and lights. Ensure that all electrical components are in good condition and that lights are functioning correctly for safe operation, especially if working in low-light conditions.

9. Grease Points: Lubricate all grease points on the baler according to the manufacturer's recommendations. Pay special attention to bearings, pivot points, and moving parts to ensure smooth operation and extend the lifespan of the equipment.

10. Documentation: Review the operator's manual and any relevant safety guidelines or procedures provided by the manufacturer. Familiarize yourself with the proper operation of the baler and any specific maintenance requirements or precautions.

Connecting a baler to a tractor involves several steps to ensure proper alignment, functionality, and safety. Here's a general guide on how to connect a baler to a tractor:

1. Positioning the Baler: Park the baler on level ground in close proximity to the tractor, ensuring there is enough space to manoeuvre the tractor and attach the baler safely.

2. Aligning the Hitch: Position the tractor directly in front of the baler's hitch mechanism. Ensure that the tractor's hitch is at the appropriate height to align with the hitch on the baler.

3. Lowering the Hitch: Lower the tractor's hitch mechanism to match the height of the baler's hitch. This may require adjusting the tractor's hydraulic lift system or using manual controls, depending on the tractor model.

4. Engaging the Hitch: Back the tractor slowly towards the baler, aiming to align the tractor's hitch with the hitch on the baler. Once aligned, lower the tractor's hitch onto the baler's hitch,

ensuring a secure connection.

5. Securing the Hitch: Once the hitches are engaged, secure them together using locking pins or other fastening mechanisms to prevent accidental disconnection during operation.

6. Connecting Hydraulic Lines: If the baler requires hydraulic power from the tractor, connect the hydraulic lines from the baler to the tractor's hydraulic system. Ensure that the connections are tight and secure to prevent leaks.

7. Testing the Connection: Before operating the baler, test the connection between the tractor and baler by applying slight forward pressure to ensure that the hitches are securely engaged and the hydraulic lines are functioning properly.

8. Adjusting the PTO Shaft: If the baler is equipped with a power take-off (PTO) shaft, align the splines on the shaft with the PTO coupling on the tractor. Carefully connect the PTO shaft, ensuring that it engages fully and locks into place.

9. Securing Safety Chains: Some balers may have safety chains or cables that need to be attached to the tractor to provide additional security in case of hitch failure. Attach these chains to designated anchor points on the tractor and baler.

10. Performing a Final Check: Once all connections are made, perform a final visual inspection to ensure that everything is properly connected and secured. Double-check that all fasteners are tight, hydraulic lines are connected correctly, and safety features are engaged.

Using a baler involves several steps to efficiently and safely compress and package agricultural materials into bales for storage, transport, or further processing. Here's a guide on how to use a baler:

Preparation: Position the baler on level ground in an open area free from obstacles. Ensure that the baler is properly connected to a compatible tractor with the necessary power source (such as a power take-off or PTO) and hydraulic connections. Review the manufacturer's operating manual and safety instructions before starting.

Adjustments: Set the baler's settings according to the type of material being baled, such as hay, straw, or silage, and the desired bale size and density. Adjust the pickup height to ensure proper gathering of the material from the ground into the baler.

Material Collection: Drive the tractor and baler over the area where the material (such as hay or straw) is located. The pickup mechanism of the baler collects the material from the ground and feeds it into the bale chamber.

Bale Formation: As the material accumulates in the bale chamber, the baler's compression system compacts it into a dense bale. The bale size and density can be adjusted based on the baler's settings and the operator's preferences. The baler may have indicators or controls to monitor the bale formation process and adjust settings as needed.

Wrapping (if applicable): If the baler is equipped with a wrapping mechanism, it may automatically apply twine, netting, or plastic film to secure the bale. Follow the manufacturer's instructions for operating the wrapping mechanism, including adjusting tension and overlap settings.

Ejection: Once the bale reaches the desired size and density, the baler's ejection system releases the bale from the chamber. Some balers may have a hydraulic or mechanical system to push the bale out, while others may rely on gravity.

Figure 81: Ejecting a bale from a Lely Welger RP 425 round baler. milo bostock, CC BY 2.0, via Wikimedia Commons.

Storage or Transport: After ejection, move the tractor and baler to a suitable location for storing or transporting the bales. Stack the bales neatly to maximize space efficiency and minimize the risk of damage during storage or transport.

Maintenance: After using the baler, perform routine maintenance tasks such as cleaning, greasing moving parts, and inspecting for any signs of wear or damage. Store the baler in a dry, sheltered area when not in use to protect it from the elements and extend its lifespan.

Using a bale wrapper involves several steps to effectively wrap bales of forage, such as hay or silage, with plastic film to create an airtight seal. Here's a guide on how to use a bale wrapper:

Preparation: Position the bale wrapper on level ground in a suitable location near the bales to be wrapped. Ensure that the bale wrapper is connected to a compatible tractor with the necessary power source (such as a power take-off or PTO) and hydraulic connections. Review the manufacturer's operating manual and safety instructions before starting.

Adjustments: Set the bale wrapper's settings according to the type of material being wrapped (hay, straw, or silage) and the desired wrapping configuration (number of layers, overlap, etc.). Adjust the height and width of the wrapping arms or platform to accommodate the size and shape of the bales.

Loading the Bales: Load the bales to be wrapped onto the bale wrapper's wrapping platform or arms using a bale spike or loader attachment on the tractor. Position the bales evenly spaced and aligned on the platform or arms to ensure uniform wrapping.

Wrapping Process: Start the wrapping process according to the manufacturer's instructions. This may involve activating the wrapping mechanism manually or through automated controls. The bale wrapper dispenses plastic film from a roll or pre-cut sheets and applies it to the bales as they rotate on the wrapping platform or arms. Ensure that the plastic film is applied evenly and securely around the bales, with sufficient overlap and tension to create an airtight seal.

Figure 82: Wrapping a bale with a tractor-pulled self-loading bale wrapper Taarup 7517 by Kverneland. Tumi-1983, CC BY-SA 3.0, via Wikimedia Commons.

Adjustments (if applicable): Monitor the wrapping process and make any necessary adjustments to the wrapping settings, such as tension, overlap, or film thickness, to ensure optimal wrapping quality.

Completion: Once the wrapping process is complete, stop the bale wrapper and disengage the wrapping mechanism. Remove the wrapped bales from the wrapping platform or arms using a bale spike or loader attachment on the tractor.

Storage or Transport: Move the wrapped bales to a suitable location for storage or transport, such as a storage shed, stack yard, or feeding area. Stack the wrapped bales neatly to maximize space efficiency and minimize the risk of damage during storage or transport.

Maintenance: After using the bale wrapper, perform routine maintenance tasks such as cleaning, greasing moving parts, and inspecting for any signs of wear or damage. Store the bale wrapper in a dry, sheltered area when not in use to protect it from the elements and extend its lifespan.

Completing Baler Operations

Finishing up baler operations, cleaning, and maintaining the baler, as well as proper storage, are crucial steps to ensure the equipment's longevity and efficient performance. Here's a detailed guide on how to complete these tasks:

Finishing Up Baler Operations:

1. Once all the bales have been produced, stop the tractor engine and disengage any hydraulic connections.
2. Lower the pickup and any moving parts to their resting positions.
3. Turn off any baler-specific systems or mechanisms, such as the knotter or tying mechanism.
4. Disconnect the baler from the tractor, ensuring all connections are safely released.
5. Inspect the surrounding area for any leftover debris or materials and remove them to prevent potential hazards for future operations.

Cleaning the Baler:

1. Start by removing any remaining materials or debris from the bale chamber, pickup, and other relevant components.
2. Use compressed air or a brush to clean hard-to-reach areas, ensuring all residues are removed.

3. Check and clean the baler's hydraulic system, ensuring there are no leaks or blockages.

4. Inspect the knives or cutting mechanisms and clean them if necessary to prevent buildup.

5. Lubricate moving parts as per the manufacturer's recommendations to maintain smooth operation.

6. Wipe down the exterior of the baler to remove any dirt or grime.

Maintenance:

1. Inspect all components for signs of wear, damage, or corrosion, and address any issues promptly.

2. Check the tension of belts, chains, and other drive components, adjusting as needed to ensure proper operation.

3. Grease bearings, pivot points, and other moving parts according to the manufacturer's recommendations.

4. Verify that safety features, such as guards and shields, are intact and functioning correctly.

5. Perform any scheduled maintenance tasks outlined in the manufacturer's manual, such as changing filters or fluids.

Storage:

1. Choose a dry, sheltered location for storing the baler to protect it from the elements.

2. If possible, store the baler indoors to minimize exposure to moisture and harsh weather conditions.

3. Cover the baler with a tarp or protective covering to further shield it from dust and debris.

4. Ensure the storage area is secure to prevent unauthorized access and potential theft.

5. If storing the baler for an extended period, consider elevating it off the ground to prevent rust and corrosion.

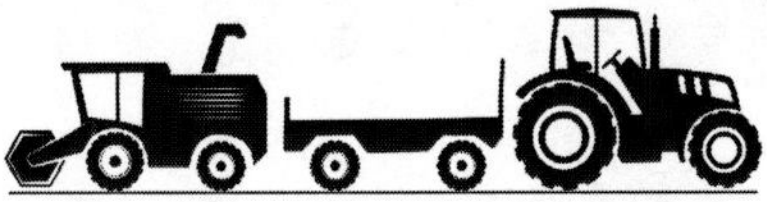

Chapter Eleven

Mowers and Slashers

An agricultural mower and slasher are types of machinery used in farming and landscaping for cutting grass, weeds, and other vegetation. While they serve a similar purpose, they differ in design and function:

1. Agricultural Mower: Agricultural mowers are machines designed to cut grass and other vegetation in fields, pastures, orchards, and along roadsides. They come in various types, including reel mowers, sickle bar mowers, rotary mowers, and flail mowers. Reel mowers use a series of blades mounted on a rotating cylinder to cut grass cleanly, while sickle bar mowers have a reciprocating blade that moves back and forth to cut vegetation. Rotary mowers use spinning blades attached to a horizontal axis to chop up grass and debris, and flail mowers have chains or blades that spin rapidly to cut vegetation into small pieces. Agricultural mowers are typically mounted on tractors or other vehicles and can be either rear-mounted or side-mounted.

2. Slasher: A slasher, also known as a bush hog or brush cutter, is a heavy-duty mower designed for cutting thick brush, saplings,

and overgrown vegetation. Slashers are often used for clearing land, maintaining fence lines, and managing rough terrain. They typically have a large, heavy-duty rotary blade mounted on a powerful motor and are designed to handle tough, woody vegetation that may be too much for standard agricultural mowers. Slashers are usually towed behind a tractor or other vehicle and are capable of cutting through dense vegetation with ease.

Both agricultural mowers and slashers play important roles in land management and agricultural operations, helping farmers and landowners maintain fields, pastures, and other outdoor spaces. The choice between them depends on the specific requirements of the task at hand, such as the type and thickness of vegetation to be cut and the terrain to be traversed.

Figure 83: Massey Ferguson 590 tractor with a rear-mounted disc mower. kallerna, CC BY-SA 4.0, via Wikimedia Commons.

The different types of mowers can be characterised as:

1. **Reel Mowers**:

   - **Design**: Reel mowers consist of a series of blades mounted on a rotating cylinder (reel) that spins perpendicular to the ground. The blades cut the grass against a stationary cutting bar, resulting in a scissor-like action.

- **Functionality**: As the reel rotates, it captures and cuts the grass blades against the cutting bar with a shearing action, producing a clean and precise cut. Reel mowers are often preferred for maintaining fine turf on golf courses, sports fields, and formal lawns due to their ability to deliver a manicured finish.
- **Advantages**: They provide a clean and even cut, are environmentally friendly (no emissions), and promote healthy turf growth by minimizing stress on the grass.
- **Disadvantages**: They are generally less effective on tall or tough grasses, require frequent adjustments and maintenance, and are usually less efficient for large areas compared to other types of mowers.

2. **Sickle Bar Mowers**:
    - **Design**: Sickle bar mowers feature a long bar with small, serrated blades attached along its length. The bar moves back and forth (reciprocates) horizontally, similar to a hand-held sickle, to cut vegetation.
    - **Functionality**: The reciprocating motion of the bar allows the serrated blades to slice through grass, weeds, and other vegetation. Sickle bar mowers are efficient at cutting tall and dense vegetation and are often used in agricultural settings for haymaking and pasture maintenance.
    - **Advantages**: They are effective for cutting tall and thick

vegetation, have a wide cutting swath, and are relatively low maintenance.

- **Disadvantages**: They may not provide as clean a cut as reel mowers, are less suitable for maintaining formal lawns, and may struggle with uneven terrain.

3. **Rotary Mowers**:

- **Design**: Rotary mowers feature a horizontal blade or blades mounted on a spindle underneath the mower deck. The blade spins rapidly around a vertical axis to cut grass and vegetation.
- **Functionality**: The spinning blade creates a powerful cutting action, effectively chopping up grass, weeds, and brush as it passes over them. Rotary mowers are versatile and widely used for maintaining lawns, fields, and rough terrain.
- **Advantages**: They are versatile and efficient for cutting a variety of grasses and vegetation, suitable for different terrains, and relatively easy to use and maintain.
- **Disadvantages**: They may produce a less uniform cut compared to reel mowers, can generate more noise and vibration, and may require more frequent blade sharpening.

4. **Flail Mowers**:

- **Design**: Flail mowers consist of a rotating drum or shaft with small, hinged blades (flails) attached. The flails swing freely as the drum rotates, striking vegetation at high speed to cut it.
- **Functionality**: Flail mowers are capable of cutting through

tough, thick vegetation, including woody brush and saplings. The swinging action of the flails allows them to handle uneven terrain and obstacles more effectively than other mower types.

- **Advantages**: They excel at cutting dense and rough vegetation, are highly durable and resistant to damage from rocks and debris, and can be used in various environments.
- **Disadvantages**: They may produce a rougher cut compared to reel mowers, are generally heavier and more expensive, and may require more power to operate efficiently.

Each type of mower has its own unique characteristics, making it suitable for different applications and environments. The choice of mower depends on factors such as the type of vegetation to be cut, the desired cut quality, the terrain, and the operator's preferences and requirements.

Figure 84: Deutz-Fahr Agrotron tractor with a Pöttinger Novadisc 350 rear-mounted disc mower (304 mm working width) and a Pöttinger front mower. Roland zh, CC BY-SA 3.0 , via Wikimedia Commons.

Figure 85: Flail mower drum. Alupus, CC BY-SA 3.0, via Wikimedia Commons.

A mower conditioner is a type of agricultural machinery used for cutting and conditioning hay or other forage crops before baling or storage. It combines the functions of a mower and a conditioning implement into a single machine, streamlining the haymaking process.

Figure 86: A John Deere tractor and a John Deere 1207 mower-conditioner. liz west (Muffet), CC BY 2.0, via Wikimedia Commons.

The primary function of a mower conditioner is to cut standing crops such as grass or alfalfa and condition the cut material to promote faster and more efficient drying. Conditioning involves crushing or crimping the stems of the forage crop to accelerate moisture release and reduce drying time. This process helps preserve the nutritional quality of the forage by minimizing the risk of mold or spoilage.

Mower conditioners typically consist of cutting components, such as blades or discs, for mowing the crop, and conditioning components, such as rollers or flails, for conditioning the cut material. The conditioning mechanism may vary depending on the model of the machine but often involves crushing or crimping the stems between rotating rollers or flails.

Mower conditioners come in various sizes and configurations to suit different farm sizes, terrain types, and crop conditions. They can be mounted on tractors or operated as self-propelled units, offering flexibility and efficiency in haymaking operations.

Overall, mower conditioners play a crucial role in modern haymaking practices, enabling farmers to efficiently harvest, condition, and preserve forage crops for use as livestock feed or bedding.

Selecting the appropriate mower, slasher, or mower conditioner for a specific task entails a thorough consideration of various factors encompassing the type of vegetation, terrain, desired outcomes, and available equipment options. Selection of the best suited equipment for the task includes considering:

- Type of Vegetation: When contemplating the type and density of vegetation to be mowed or slashed, it's crucial to match the equipment accordingly. Light grasses and weeds may be adequately managed with a slasher, while thicker, heavier vegetation like hay or alfalfa may necessitate the enhanced capabilities of a mower conditioner.

- Terrain: The terrain where the mowing or slashing will occur significantly influences equipment selection. While a standard mower or slasher might suffice for flat, even terrain, uneven or rough landscapes may require a mower conditioner equipped with adjustable cutting height and terrain-following features to ensure uniform cutting and conditioning.

- Cutting Width: Determining the appropriate cutting width based on the size of the area to be mowed or slashed is essential for optimizing efficiency and minimizing task duration. Mower conditioners typically offer wider cutting widths compared to slashers, which can expedite operations on larger fields.

- Conditioning Needs: Assessing the necessity for conditioning in your application is paramount. If you aim to expedite drying and enhance forage quality, a mower conditioner with conditioning rollers or flails may be preferred. Conversely, if conditioning is unnecessary, a standard mower or slasher may suffice.

- Tractor Compatibility: Ensuring compatibility with the tractor's horsepower and hydraulic capabilities is critical to prevent inefficiencies and equipment damage. Overloading the tractor with an undersized or incompatible implement can compromise performance and durability.

- Budget: Considering budget constraints and weighing the cost-effectiveness of different options is prudent. While mower conditioners generally entail a higher initial cost than slashers or standard mowers due to their advanced features, assessing long-term benefits, such as reduced drying time and improved forage quality, is essential.

- Maintenance Requirements: Evaluating the maintenance needs of the selected equipment, including blade sharpening, lubrication, and adjustment requirements, is imperative. Opt for a mower, slasher, or mower conditioner that aligns with your maintenance capabilities and schedule to ensure sustained performance and longevity.

The hazards associated with agricultural mowing and slashing primarily revolve around the operation of machinery and the nature of the work environment. Here are some of the key hazards:

1. Contact with Moving Parts: One of the most significant hazards involves contact with the moving parts of mowing and slashing equipment. This includes blades, flails, and other cutting mechanisms. Operators risk injury if they come into contact with these moving components, which can cause cuts, lacerations, or even amputations.

2. Entanglement and Entrapment: Operators may become entangled or trapped in the machinery while it is in operation. This can occur if loose clothing, hair, or body parts are caught by

moving parts of the equipment. Entanglement and entrapment accidents can result in serious injuries or fatalities.

3. Crushing and Trapping: There is a risk of being crushed or trapped by the machinery, especially during maintenance activities or when transitioning the equipment between operational and transport positions. Operators must take precautions to prevent being caught between moving parts or pinned beneath heavy machinery.

4. Thrown Objects and Debris: Mowing and slashing equipment can eject objects and debris at high speeds, posing a risk to operators and bystanders. Rocks, branches, and other projectiles can cause serious injuries if they strike individuals in the vicinity of the equipment.

5. Overturns and Rollovers: Operating mowing and slashing equipment on uneven terrain, slopes, or banks increases the risk of overturns and rollovers. This can result in operators being thrown from the equipment or trapped beneath it, leading to injuries or fatalities.

6. Noise and Vibration: Mowing and slashing machinery often generate high levels of noise and vibration, which can pose risks to operators' hearing and musculoskeletal health over time. Prolonged exposure to loud noise and excessive vibration can cause hearing loss, fatigue, and other health issues.

7. Chemical Exposure: In some cases, mowing and slashing may involve the use of herbicides or pesticides to control vegetation. Operators risk exposure to these chemicals through inhalation, skin contact, or ingestion, leading to health problems if proper safety precautions are not followed.

8. Environmental Hazards: Operating mowing and slashing equipment in outdoor environments exposes operators to various environmental hazards, such as extreme temperatures, inclement weather, uneven terrain, and wildlife encounters. These factors can increase the risk of accidents and injuries if not properly managed.

Preparing for Mowing Operations

Preparing for agricultural mowing or slashing operations involves several important steps to ensure safety, efficiency, and effectiveness.

Assess the Area: Before beginning mowing or slashing, survey the area to be worked on. Identify any potential hazards, such as uneven terrain, obstacles, or overhead power lines. Clear the area of debris, rocks, and other objects that could be thrown by the machinery.

Check the Equipment: Inspect the mowing or slashing equipment thoroughly before use. Ensure that all components are in good working condition, including blades, flails, belts, and hydraulic systems. Check for signs of wear or damage, and make any necessary repairs or replacements.

Fuel and Fluids: Verify that the equipment has an adequate supply of fuel, oil, and hydraulic fluid. Refill or top up as needed to prevent interruptions during operation. Follow proper procedures for handling and storing fuels and fluids to minimize the risk of spills or leaks.

Safety Gear: Equip yourself with appropriate personal protective equipment (PPE), including eye protection, hearing protection, gloves, and sturdy footwear. Depending on the conditions and tasks involved, additional safety gear such as a helmet or high-visibility clothing may be necessary.

Plan the Route: Determine the most efficient route for mowing or slashing, taking into account factors such as the layout of the land, the type of vegetation to be cut, and any obstacles or hazards to avoid. Plan for safe entry and exit points, especially if working in confined or obstructed areas.

Communicate: If working as part of a team, ensure clear communication among all personnel involved in the operation. Establish signals or verbal cues to coordinate movements and actions, especially when operating heavy machinery near other workers or vehicles.

Weather Conditions: Check the weather forecast before starting work. Avoid mowing or slashing in adverse weather conditions such as heavy rain, high winds, or extreme heat. Wet or slippery conditions can increase the risk of accidents and reduce the effectiveness of the equipment.

Emergency Preparedness: Familiarize yourself with emergency procedures and protocols in case of accidents, injuries, or equipment malfunctions. Ensure that first aid supplies, communication devices, and emergency contact information are readily available on-site.

Training and Certification: Ensure that operators are properly trained and certified to operate mowing or slashing equipment safely. Provide refresher training as needed to reinforce safety protocols and best practices.

Environmental Considerations: Consider the environmental impact of mowing or slashing operations, such as soil erosion, habitat disruption, and pesticide use. Implement measures to minimize environmental damage, such as using eco-friendly equipment and techniques.

Hazards: The primary risk of injury stems from potential contact with the machine's moving components. These include active blades or flails, operating drive mechanisms, rotating power take-off (PTO) shafts, and swiftly moving blades, flails, or other attachments that may be ejected from the machine.

Additional risks associated with mowing operations involve various scenarios. There's the potential for contact with moving parts when clearing blockages. Transitioning the mower between operational and transport positions poses the risk of being trapped or crushed. Additionally, there's the danger of being crushed beneath a raised machine while replacing blades or conducting maintenance tasks. Furthermore, operators face the possibility of being struck or run over by the tractor and mower. Operating on slopes or banks increases the potential for overturns. Lastly, there's the risk of being struck by objects or debris expelled from the machine.

Control measures: Performing work on a machine while it's powered poses significant danger. The foremost safety measure is to adhere to the 'safe stop' procedure before conducting any maintenance or adjustments, including addressing blockages or other issues:

1. Engage the handbrake.
2. Ensure controls are in neutral.
3. Stop the engine.
4. Remove the key. Always allow ample run-down time for blades and cutting mechanisms before approaching or working on the machine. It's imperative that all personnel understand how to work safely. Clear instructions, information, and adequate training must be provided to employees regarding:

- The potential risks they may encounter.
- The implemented measures to control these risks.
- How to follow emergency procedures. Particular attention should be paid to the training requirements and supervision of:
- New recruits and trainees, especially young workers who are

more susceptible to accidents.

- Lone workers.

- Individuals changing roles or taking on new responsibilities.

- Health and safety representatives, who are subject to specific laws. Guarding: The power take-off shaft should be fully enclosed in a guard along its entire length, from the tractor power take-off to the power input connection on the mower. Guards must prevent contact with any hazardous part of the machine during operation. Stand-off guards should be appropriately secured, positioned both front and rear to prevent contact with the blades. Ensure all guards are correctly positioned, fitted, and secured before commencing work. Do not continue work if guards are missing or damaged. Ensure the protective skirt or other device to prevent the ejection of objects or debris is in place and in good condition. Note that other mower parts, such as conditioners, should be guarded to the same standard as the rest of the mower. General guidance on safe working practice: Ensure anyone operating a mower has received adequate instructions and training. Training courses may be available from reputable sources such as local training providers, agricultural colleges, and machinery dealers. Follow the instructions and precautions outlined in the operator's manual. Health and Safety: Do not operate a mower if there are bystanders who might be hit by broken blades or ejected debris, as these can travel considerable distances. Cease work if someone approaches during mowing activities. Exercise caution when working on steep ground, particularly during turns, especially with mounted mowers. Be vigilant for potential blockages—avoiding blockages is preferable to clearing them.

Performing prestart checks on a mower or slasher before beginning operations is crucial to ensure the equipment is in proper working condition and safe to use. This includes:

1. Visual Inspection: Conduct a visual inspection of the entire machine, checking for any signs of damage, wear, or leaks. Look for loose or missing components, such as bolts, nuts, or guards. Ensure that all safety decals and warning labels are legible and in place.

2. Fluid Levels: Check the levels of essential fluids, including fuel, engine oil, hydraulic fluid, and coolant. Top up any fluids that are low to prevent interruptions during operation. Inspect for any signs of leaks around fluid reservoirs or hoses.

3. Belts and Drives: Inspect the condition of belts, chains, and drive mechanisms. Look for signs of wear, fraying, or stretching, and replace any damaged components as necessary. Ensure that belts are properly tensioned and aligned for smooth operation.

4. Blades or Flails: Check the condition of cutting blades or flails, ensuring they are sharp, straight, and securely mounted. Replace any damaged or worn blades to maintain cutting efficiency and prevent accidents.

5. Safety Features: Test all safety features and controls, including emergency stop buttons, kill switches, and safety interlocks. Ensure that guards and shields are in place and functioning correctly to prevent contact with moving parts.

6. Brakes and Clutches: Test the brakes and clutches to ensure they are functioning properly. Check for any signs of slippage or abnormal behaviour when engaging or disengaging these components.

7. Tyres and Wheels: Inspect the condition of tyres and wheels, checking for proper inflation, tread wear, and damage. Ensure that wheel nuts are tight and secure to prevent wheel loss or detachment during operation.

8. Lights and Signals: Test all lighting and signalling systems, including headlights, taillights, indicators, and hazard lights. Ensure that lights are working correctly to maintain visibility and safety, especially when operating in low-light conditions.

9. Controls and Steering: Test the operation of all controls, levers, and steering mechanisms. Ensure that controls move freely and smoothly without binding or sticking. Check for excessive play or looseness in the steering system.

10. Operator Comfort and Ergonomics: Check the operator's seat, armrests, and foot pedals for comfort and proper adjustment. Ensure that the operator's cab or platform is clean, well-maintained, and free of obstructions for safe and comfortable operation.

Operating a Mower

Connecting a mower or slasher to a tractor typically involves utilizing a three-point hitch system, a standard method for attaching agricultural implements. The tractor's three-point hitch system comprises two lower lift arms and an upper link. Positioned at the rear of the tractor, these lower lift arms can be hydraulically raised or lowered to lift and lower the implement, while the upper link enhances stability and controls the angle of the implement.

The mower or slasher is engineered with attachment points that correspond to the tractor's three-point hitch. These attachment points, typically located near the rear of the implement, may comprise mounting pins or brackets, ensuring alignment and secure connection.

To initiate the connection, the tractor is maneuverer behind the mower or slasher, aligning the three-point hitch arms with the attachment points on the implement. Fine adjustments may be necessary to achieve proper alignment, allowing for optimal functionality.

Once aligned, the operator lowers the tractor's three-point hitch arms to engage with the attachment points on the mower or slasher, establishing a secure connection between the two.

Depending on the specific design of the implement, additional securing mechanisms, such as locking pins, bolts, or fasteners, may be employed to reinforce the connection and prevent detachment during operation.

For implements requiring hydraulic power, hydraulic hoses are connected from the tractor to the implement. These hoses supply hydraulic fluid necessary for operating hydraulic cylinders or motors on the implement.

Figure 87: Flail mower/mulcher connected to tractor. Alupus, CC BY-SA 3.0, via Wikimedia Commons.

Prior to commencing operations, the operator should conduct a thorough inspection to ensure all connections are secure and correctly aligned. Additionally, any safety guards or shields on the implement should be in place and functioning correctly to mitigate the risk of accidents.

Operating an agricultural mower demands careful adherence to safety protocols, proper handling of equipment, and proficient mowing techniques. Below is a stepwise guide detailing how to operate an agricultural mower:

Preparation: Conduct a thorough pre-operation inspection of the mower, scrutinizing for any indications of damage, wear, or leaks. Ensure that all components, including blades, belts, and safety guards, are not only in good condition but also securely fastened. Check the fluid levels, encompassing fuel, engine oil, hydraulic fluid, and coolant, and replenish them as necessary to avoid interruptions during operation. Don appropriate personal protective equipment (PPE), such as eye protection, hearing protection, gloves, and sturdy footwear, to mitigate the risk of injury.

Start-Up: Assume the operator's position and fasten the seatbelt if the tractor is equipped with one. Initiate the tractor's engine as per the manufacturer's instructions, ensuring a smooth start-up process. Activate the parking brake and ascertain that the transmission is set to neutral. Using the tractor's hydraulic controls, lower the mower deck to the desired cutting height, adjusting it according to the terrain and grass type.

Operation: Commence mowing in a straight line, initiating from the perimeter of the mowing area and progressing inward. Maintain a consistent speed, typically recommended by the manufacturer, to ensure uniform cutting and minimize turf damage. Exercise caution to avoid sudden turns or abrupt movements that could compromise control or cause damage to the mower. Remain vigilant for obstacles such as rocks, stumps, or concealed debris, manoeuvring away from them to prevent potential damage to the mower and mitigate the risk of injury. Regularly monitor the engine temperature and other gauges displayed on the tractor's dashboard, promptly addressing any anomalies or issues that arise.

Adjustments: Modify the cutting height as necessary, contingent upon the terrain, grass type, and desired outcomes. Some mowers permit on-the-go adjustments, while others may necessitate halting the tractor to make alterations. Monitor the condition of the blades vigilantly, replacing them if they become dull or damaged to uphold cutting efficiency and quality.

Completing Mowing Operations

After completing the mowing task, it is essential to follow these steps for proper closure:

- Disengage the mower deck and elevate it to its highest position using the tractor's hydraulic controls. This action prevents potential damage to the mower deck and ensures safe transportation and storage.

- Turn off the tractor's engine and disengage the parking brake to halt all operations safely.

- Conduct a thorough post-operation inspection of the mower, meticulously checking for any signs of damage, wear, or malfunction. Address any identified issues promptly to maintain the mower's performance and longevity.

- Clean the mower deck and underside meticulously to eliminate grass clippings and debris accumulation, which can lead to corrosion over time and compromise the mower's functionality. This step also aids in preventing potential safety hazards and ensures optimal performance during subsequent uses.

- Finally, store the mower in a dry, secure location away from the elements and potential hazards. Proper storage helps preserve the mower's condition and prolong its lifespan, ensuring it remains ready for future use.

Maintenance: Adhere to the manufacturer's guidelines when replacing blades and flails. Verify that replacement parts adhere to the manufacturer's specifications. Utilize a purpose-designed prop or stand for safe maintenance beneath a mower. Avoid relying solely on tractor hydraulics. Keep in mind that even with a supported machine, heavy components may require extra support. Consider using suitable protective gloves when replacing blades and performing similar maintenance tasks.

Chapter Twelve

Spreaders – Fertilizer and Manure

A fertilizer or manure spreader is a piece of agricultural equipment used to evenly distribute fertilizers, manure, or other soil amendments onto fields or crops. These spreaders are designed to efficiently and accurately apply nutrients or organic matter to the soil, promoting healthy plant growth and maximizing crop yields.

Farmers spread fertilizers and manure for various reasons, all centred around improving soil fertility and fostering healthy plant growth. These practices are fundamental in modern agriculture, and here are the primary motivations behind them:

Nutrient Supply: Fertilizers and manure serve as rich sources of essential nutrients crucial for plant growth, including nitrogen, phosphorus, potassium, and secondary micronutrients such as calcium, magnesium, and sulphur. By spreading fertilizers and manure, farmers ensure the soil receives these vital nutrients, promoting optimal growth and productivity in crops.

Soil Improvement: Fertilizers and manure play a crucial role in enhancing soil quality by improving its physical, chemical, and biological properties. They contribute to increasing soil organic matter content, enhancing soil structure and texture, improving water retention and drainage, and fostering beneficial microbial activity. These improvements create a conducive environment for root development and nutrient absorption, ultimately leading to better crop yields and quality.

Crop Yield Enhancement: Maximizing crop yields is a primary objective of spreading fertilizers and manure. By supplying essential nutrients to the soil, these substances stimulate plant growth, enhance photosynthesis, and increase biomass production. Consequently, farmers achieve higher yields of grains, fruits, vegetables, or other agricultural products, crucial for meeting food demands and ensuring farm profitability.

Nutrient Balance: Continuous crop cultivation gradually depletes soil nutrients, resulting in deficiencies and reduced yields over time. Spreading fertilizers and manure helps maintain a balance of nutrients in the soil by replenishing lost elements and addressing deficiencies. Farmers often conduct soil tests to assess nutrient levels, allowing them to adjust fertilizer and manure applications accordingly to meet crop requirements and sustain soil fertility in the long run.

Environmental Stewardship: Responsible nutrient management, including the judicious use of fertilizers and manure, is vital for sustainable agriculture and environmental preservation. Applying these substances at recommended rates and timings helps minimize nutrient runoff and leaching, which can contribute to water pollution, eutrophication, and other environmental issues. By adopting sustainable nutrient management practices, farmers safeguard water quality, preserve soil health, and protect ecosystems for future generations.

Spreading fertilizers and manure is a cornerstone practice in modern agriculture, aimed at enhancing soil fertility, increasing crop yields, and promoting sustainable land management. By diligently managing nutrient inputs and adhering to best practices, farmers can optimize agricultural productivity while mitigating adverse environmental impacts.

The spreader typically consists of a hopper or container that holds the fertilizer or manure, a mechanism for breaking up and distributing the material, and a spreading mechanism that evenly disperses the material over the field. Some spreaders are designed to be towed behind a tractor, while others may be mounted directly onto the tractor or other agricultural equipment.

Fertilizer or manure spreaders come in various sizes and configurations to suit different farming operations and field sizes. They may be equipped with adjustable settings to control the rate and pattern of material distribution, allowing farmers to tailor the application to specific crop requirements and soil conditions.

Figure 88: Rolland manure spreader V2-160, Deutz-Fahr Agrotron tractor. werktuigendagen, CC BY-SA 2.0, via Wikimedia Commons.

Using a fertilizer or manure spreader offers several benefits, including:

1. Improved Nutrient Distribution: By evenly spreading fertilizers or manure across the field, spreaders ensure that nutrients are distributed uniformly, helping to prevent over-application in some areas and under-application in others.

2. Increased Efficiency: Spreaders allow farmers to cover large areas of land quickly and efficiently, reducing the time and labour required for manual application methods.

3. Cost Savings: By accurately applying nutrients or organic matter only where needed, spreaders help farmers optimize their use of fertilizers and manure, reducing waste and minimizing input costs.

4. Enhanced Crop Yield and Quality: Properly applied fertilizers or manure can improve soil fertility, leading to healthier plants, increased crop yields, and improved crop quality.

Overall, fertilizer or manure spreaders are essential tools for modern agriculture, enabling farmers to effectively manage soil fertility, promote sustainable farming practices, and maximize productivity.

Tractor-pulled and dedicated or self-propelled fertilizer/manure spreaders differ primarily in their design, functionality, and intended use. Here's an explanation of the key differences between the two:

1. **Design and Construction**:

    - Tractor-pulled spreaders are designed to be towed behind a tractor, typically using a hitch mechanism. They are often connected to the tractor's power take-off (PTO) to utilize the tractor's engine power for operation. These spreaders may vary in size and configuration but are generally built to be compatible with various tractors.

- Dedicated fertilizer/manure spreaders, on the other hand, are standalone machines specifically designed for spreading fertilizers or manure. They are typically mounted on trucks, trailers, or other vehicles and may have their own power source, such as an engine or hydraulic system, for operation. Dedicated spreaders are purpose-built for spreading specific types of materials and may offer specialized features for efficient application.

2. **Versatility and Flexibility**:

- Tractor-pulled spreaders offer greater versatility as they can be easily attached to and detached from different tractors, allowing farmers to use the same spreader for various tasks and with different tractor models. They are well-suited for small to medium-sized farms with diverse needs.

- Dedicated fertilizer/manure spreaders are designed for specific applications and materials, offering less flexibility compared to tractor-pulled spreaders. While they may excel at spreading certain types of fertilizers or manure, they may not be suitable for other tasks or materials without modifications.

3. **Capacity and Efficiency**:

- Tractor-pulled spreaders typically have smaller capacities compared to dedicated spreaders, as they rely on the tractor's towing capacity and may need to be compatible with different tractor sizes. While they may be less efficient in terms of capacity, tractor-pulled spreaders can still offer adequate performance for many agricultural operations.

  - Dedicated fertilizer/manure spreaders are often larger and more efficient in terms of capacity, allowing for higher throughput and reduced downtime. They are designed to handle specific materials with optimal efficiency and may offer features such as larger hoppers, faster spreading rates, and more advanced control systems.

4. **Cost and Investment**:

  - Tractor-pulled spreaders generally have lower upfront costs compared to dedicated spreaders, as they leverage existing tractor equipment and infrastructure. They are a cost-effective option for smaller farms or those with limited budgets.

  - Dedicated fertilizer/manure spreaders tend to have higher upfront costs due to their specialized design and capabilities. While they may require a larger initial investment, they can offer increased efficiency and productivity, particularly for larger-scale farming operations.

In summary, tractor-pulled spreaders offer versatility and flexibility, making them suitable for a wide range of agricultural tasks and farm sizes. Dedicated fertilizer/manure spreaders, on the other hand, are purpose-built for specific applications and materials, offering higher capacity and efficiency but with a higher initial investment. The choice between the two depends on factors such as farm size, budget, and the specific requirements of the task at hand.

Figure 89: Lite-Trac Agri-Spread lime and fertiliser spreader (Self-propelled or dedicated). Lite-Trac, CC BY-SA 3.0, via Wikimedia Commons.

Fertilizer and manure spreaders come in various types, each designed to suit specific applications, field sizes, and operational preferences. Here are the different types of fertilizer and manure spreaders commonly used in agriculture:

1. **Broadcast Spreaders**:

    - Broadcast spreaders are designed to evenly distribute fertilizers or manure over a wide area in a fan-like pattern. They are commonly used for broadcasting granular fertilizers, seeds, lime, or other dry materials. Broadcast spreaders can be either tow-behind or mounted directly onto tractors, and they typically utilize a spinning disk or agitator to disperse the material.

2. **Drop Spreaders**:
    - Drop spreaders deposit fertilizers or manure in a precise, controlled manner directly underneath the spreader. They are ideal for accurately applying materials in narrow rows or confined spaces, such as garden beds or smaller fields. Drop spreaders are often used for granular fertilizers or seed broadcasting and are available in both tow-behind and walk-behind configurations.
3. **Pendulum Spreaders**:
    - Pendulum spreaders utilize a swinging pendulum mechanism to distribute fertilizers or manure laterally across the field. This type of spreader offers excellent accuracy and uniformity, making it suitable for large-scale agricultural applications. Pendulum spreaders are commonly towed behind tractors and are preferred for spreading granular fertilizers or lime.
4. **Spinner Spreaders**:
    - Spinner spreaders, also known as centrifugal spreaders, use a spinning disc or impeller to propel fertilizers or manure outward in a circular pattern. They are versatile and can handle a wide range of materials, including granular, powdered, or pelletized substances. Spinner spreaders are available in tow-behind, mounted, or handheld configurations, making them suitable for various farm sizes and terrain types.
5. **Liquid Manure Spreaders**:
    - Liquid manure spreaders are specifically designed to handle and distribute liquid manure or slurry onto fields. They

typically consist of a tank for holding the liquid, a pumping system to transfer the material, and a distribution mechanism such as hoses or booms to apply the liquid evenly. Liquid manure spreaders are commonly used in livestock operations and are often mounted on trucks or trailers for mobility.

6. **Drag Chain Spreaders**:
   - Drag chain spreaders utilize a conveyor belt or chain to transport and distribute solid manure or compost onto the field. They are well-suited for handling thicker, heavier materials and can be particularly useful for organic farming operations. Drag chain spreaders are available in tow-behind or mounted configurations and offer adjustable spreading widths to accommodate different field sizes.

By understanding the characteristics and capabilities of each type of fertilizer and manure spreader, farmers can select the most suitable equipment for their specific needs, ensuring efficient and effective nutrient management in agricultural operations.

Broadcast spreaders function by dispersing fertilizers, seeds, lime, or other dry materials uniformly over a wide area, creating a fan-like pattern. These spreaders are particularly favoured for their ability to efficiently cover extensive land surfaces, rendering them indispensable in agricultural and landscaping endeavours.

The operation of broadcast spreaders can be broken down into several key steps.

Hopper Loading: Initially, the materials intended for spreading, such as granular fertilizers or seeds, are loaded into the spreader's hopper. The hopper's capacity typically varies, ranging from a few hundred pounds to several thousand pounds, contingent on the spreader's size.

Distribution Mechanism: Within the hopper, a distribution mechanism is situated to ensure a consistent flow of material toward the spreading apparatus. This component might comprise an agitator, conveyor belt, or auger, facilitating the movement of material toward the spreading mechanism.

Figure 90: Electric twin broadcaster APV ZS 200 M2 mounted on a Deutz-Fahr Agrotron. Apvtpgesmbh, CC BY-SA 3.0, via Wikimedia Commons.

Spreading Mechanism: The spreading mechanism is tasked with dispersing the material evenly across the ground. Typically, broadcast spreaders employ a spinning disk or impeller located at the hopper's bottom. As the disk rotates, it projects the material outward in a fan-like pattern, ensuring uniform coverage.

Adjustable Settings: Many broadcast spreaders are equipped with adjustable settings that enable operators to regulate the rate and pattern of material distribution. These settings empower operators to customize the spread width and density to suit the specific demands of the task and the type of material being spread.

Tow-Behind or Mounted: Broadcast spreaders are available in two primary configurations: tow-behind and mounted. Tow-behind spreaders are affixed to a hitch at the rear of a tractor or ATV and are towed behind the vehicle during operation. In contrast, mounted spreaders are directly attached to the tractor or other equipment, typically utilizing a three-point hitch.

Figure 91: Tow-behind broadcast spreader B82. Gmihail at Serbian Wikipedia, CC BY-SA 3.0 RS, via Wikimedia Commons.

Operation: To operate a broadcast spreader, the operator fills the hopper with the desired material and adjusts the settings for spread width and density. Subsequently, the operator drives the vehicle over the area slated for treatment. As the vehicle advances, the spinning disk or impeller disperses the material evenly over the ground, effectively covering a broad area with each pass.

Broadcast spreaders serve as efficient and effective tools for disseminating dry materials over expansive land surfaces, catering to the needs of agricultural, landscaping, and turf management applications.

Drop spreaders are designed to deposit fertilizers or manure with precision and control directly underneath the spreader. Unlike broadcast spreaders that distribute materials in a fan-like pattern, drop spreaders offer targeted application, making them ideal for narrow rows or confined spaces such as garden beds or smaller fields.

The operation of drop spreaders typically involves several key steps:

Loading the Hopper: Materials such as granular fertilizers or seeds are loaded into the hopper of the drop spreader. The hopper is engineered with a narrow opening to ensure accurate placement of the material.

Distribution Mechanism: Inside the hopper, a distribution mechanism regulates the flow of material downward onto the ground. This mechanism, which may consist of gates or plates, opens and closes to control the release of the material.

Drop Pattern: As the drop spreader advances, the distribution mechanism releases the material directly underneath the spreader in a controlled manner. This results in a narrow and concentrated drop pattern, ensuring precise placement of the material without wastage or over-spreading.

Adjustable Settings: Many drop spreaders are equipped with adjustable settings to control the width of the drop pattern and the rate of material application. These settings enable operators to tailor the spread width and density to meet the specific requirements of the task and the type of material being spread.

Tow-Behind or Walk-Behind Configurations: Drop spreaders are available in both tow-behind and walk-behind configurations. Tow-behind drop spreaders are connected to a hitch at the rear of a tractor or ATV and are towed behind the vehicle during operation. Walk-behind drop spreaders are operated manually by pushing them forward.

Drop spreaders excel in providing precise applications of granular fertilizers or seeds, particularly in areas where accuracy and control are crucial. Their ability to deliver materials directly underneath the spreader makes them essential tools for gardeners, landscapers, and farmers, especially in smaller or more restricted spaces.

Pendulum spreaders employ a swinging pendulum mechanism to distribute fertilizers or manure laterally across agricultural fields. Unlike broadcast or drop spreaders that disperse materials in fan-like patterns or directly beneath the spreader, pendulum spreaders ensure excellent accuracy and uniformity across large-scale applications by offering lateral distribution.

The operation of pendulum spreaders typically includes:

- Loading the Hopper: Granular fertilizers or lime, among other materials, are loaded into the hopper of the pendulum spreader, much like other spreader types. The hopper is specially designed to securely hold the materials and prevent spillage during operation.

- Pendulum Mechanism: A pivotal component of pendulum spreaders is the swinging pendulum mechanism located beneath the hopper. As the spreader advances, the pendulum swings back and forth, facilitating the lateral distribution of material across the field.

- Distribution Pattern: The swinging motion of the pendulum ensures an even distribution of material over a wide area. This mechanism guarantees excellent accuracy and uniformity, as the material is consistently spread across the field.

- Adjustable Settings: Many pendulum spreaders are equipped with adjustable settings to regulate the rate and width of material distribution. These settings empower operators to customize the spread width and density based on the specific task requirements and material types.

- Towed Behind Tractors: Pendulum spreaders are typically towed behind tractors using hitch mechanisms. They are engineered to be compatible with various tractor sizes and configurations, rendering them versatile tools for agricultural operations.

Pendulum spreaders are favoured for spreading granular fertilizers or lime due to their capacity to provide accurate and uniform distribution over expansive areas. Their lateral spreading mechanism ensures efficient coverage while minimizing the risk of over- or under-application. Consequently, pendulum spreaders play a pivotal role in modern agriculture by contributing to field fertilization and optimizing crop yields.

Spinner spreaders, also referred to as centrifugal spreaders, operate by utilizing a spinning disc or impeller to propel fertilizers or manure outward in a circular pattern. Unlike other spreader types that offer lateral or drop distribution, spinner spreaders disperse materials in a radial fashion, covering a wide area with each rotation. These spreaders are known for their versatility and ability to handle a diverse range of materials, including granular, powdered, or pelletized substances.

Spinner spreaders typically operate by:

1. Loading the Hopper: Similar to other spreader types, materials such as fertilizers or manure are loaded into the hopper of the spinner spreader. The hopper is designed to contain and distribute the material effectively during operation.

2. Spinning Disc or Impeller: The key component of a spinner spreader is the spinning disc or impeller located at the bottom of the hopper. As the spreader moves forward, the disc or impeller rotates rapidly, creating centrifugal force that propels the material outward in a circular pattern.

3. Distribution Pattern: The spinning disc or impeller disperses the material in a radial pattern around the spreader, covering a wide

area with each rotation. This distribution pattern ensures thorough coverage and uniform application of the material across the field.

4. Adjustable Settings: Many spinner spreaders are equipped with adjustable settings to control the rate and width of material distribution. Operators can customize these settings based on the specific requirements of the task and the type of material being spread.

5. Configurations: Spinner spreaders are available in various configurations, including tow-behind, mounted, or handheld models. Tow-behind spreaders are attached to a hitch at the rear of a tractor or ATV and are towed behind the vehicle during operation. Mounted spreaders are directly attached to the tractor or other equipment, typically using a three-point hitch. Handheld models are operated manually by hand and are suitable for smaller-scale applications or spot treatments.

Spinner spreaders are valued for their versatility and efficiency in handling different types of materials and terrain. Their ability to distribute materials in a radial pattern makes them suitable for various farm sizes and terrain types, ranging from small gardens to large agricultural fields.

Liquid manure spreaders are specialized machines designed to handle and distribute liquid manure or slurry onto fields. They are essential tools in agricultural operations, particularly in livestock farming where manure management is crucial for soil fertility and crop production.

Figure 92: Fendt 820 Vario tractor with a Joskin liquid manure trailer. werktuigendagen, CC BY-SA 2.0, via Wikimedia Commons.

Liquid manure spreaders typically work as follows:

1. Tank for Holding Liquid: Liquid manure spreaders feature a large tank or reservoir where the liquid manure is stored. The tank is typically constructed from durable materials such as steel or polyethylene to withstand the corrosive nature of manure.

2. Pumping System: A pumping system is used to transfer the liquid manure from the storage tank to the distribution mechanism. This pumping system may consist of a hydraulic or PTO-driven pump, which creates pressure to move the liquid through the system.

3. Distribution Mechanism: Liquid manure spreaders are equipped with a distribution mechanism to apply the liquid manure evenly across the field. This mechanism can vary depending on the specific design of the spreader but often includes

hoses, booms, or nozzles mounted on the rear of the spreader.

4. Application Process: During operation, the pumping system draws liquid manure from the tank and transfers it to the distribution mechanism. The liquid manure is then sprayed or injected onto the field through the hoses, booms, or nozzles. The distribution mechanism is typically adjustable, allowing operators to control the rate and pattern of application to ensure even coverage.

5. Mobility: Liquid manure spreaders are often mounted on trucks or trailers for mobility. This allows them to be easily transported between fields and maneuvered around the farm. Some liquid manure spreaders may also be mounted on tractors or other agricultural machinery for added versatility.

Figure 93: Vredo VT 3936 self-propelled trac with slurry tank and injector in action. Joost J. Bakker from IJmuiden, CC BY 2.0, via Wikimedia Commons.

Liquid manure spreaders play a vital role in nutrient management and soil health by recycling organic matter from livestock operations back into the soil as fertilizer. By efficiently applying liquid manure to fields, these spreaders help improve soil fertility, enhance crop yields, and reduce environmental pollution from excess nutrients.

Drag chain spreaders function by utilizing a conveyor belt or chain to transport and distribute solid manure or compost onto the field. These spreaders are designed to handle thicker, heavier materials compared to other types of spreaders and are particularly suitable for organic farming operations where solid organic matter needs to be evenly distributed across fields. Here's how drag chain spreaders typically work:

1. Conveyor Belt or Chain: The key component of a drag chain spreader is the conveyor belt or chain system. This system is responsible for moving solid manure or compost from the storage compartment of the spreader to the distribution mechanism.

2. Loading the Storage Compartment: Solid manure or compost is loaded into the storage compartment of the drag chain spreader. This compartment is typically located at the rear of the spreader and is designed to securely hold the material during transportation and spreading.

3. Transportation Process: As the spreader moves forward, the conveyor belt or chain system continuously transports the solid material from the storage compartment towards the rear of the spreader. The material is gradually released onto the field as it moves along the conveyor belt or chain.

4. Distribution Mechanism: Drag chain spreaders are equipped with a distribution mechanism to evenly distribute the solid material across the field. This mechanism may consist of spinning disks, beaters, or rotating paddles that help disperse the material in a wide pattern as it is released from the conveyor belt or chain.

5. Adjustable Spreading Widths: Many drag chain spreaders offer adjustable spreading widths to accommodate different field sizes and application rates. Operators can customize the spreading width according to the specific requirements of the task and the type of material being spread.

6. Tow-Behind or Mounted Configurations: Drag chain spreaders are available in tow-behind or mounted configurations. Tow-behind spreaders are attached to a hitch at the rear of a tractor or ATV and are towed behind the vehicle during operation. Mounted spreaders are directly attached to the tractor or other equipment, typically using a three-point hitch.

Overall, drag chain spreaders provide an efficient and effective means of distributing solid manure or compost onto fields, making them valuable tools for organic farming operations. Their ability to handle thicker, heavier materials and offer adjustable spreading widths makes them versatile options for a wide range of agricultural applications.

Planning and Preparing for Spreader Operations

Operating spreaders, whether they are used for fertilizers, manure, or other materials, comes with several hazards that can pose risks to both personnel and equipment. Here are some common hazards associated with spreader operations along with mitigation strategies:

1. **Contact with Moving Parts**: Spreaders often have moving parts such as conveyor belts, spinning discs, or chains which can cause serious injuries if they come into contact with operators or bystanders.

   - Mitigation: Operators should receive thorough training on safe operating procedures and be aware of all moving parts on the equipment. Safety guards and shields should be in place and properly secured to prevent accidental contact. Personal protective equipment (PPE) such as gloves, eye protection, and hearing protection should be worn at all times.

2. **Falls and Slips**: Spreader operations may take place in uneven or slippery terrain, increasing the risk of falls and slips for operators.
   - Mitigation: Operators should wear appropriate footwear with good traction to reduce the risk of slips and falls. Spreader equipment should be equipped with anti-slip surfaces or treads where operators frequently walk. Areas with potential slip hazards should be clearly marked, and operators should exercise caution when navigating such areas.
3. **Chemical Exposure**: When spreading fertilizers or chemicals, there is a risk of exposure to harmful substances through inhalation, skin contact, or ingestion.
   - Mitigation: Operators should wear appropriate PPE, including gloves, goggles, respirators, and protective clothing, when handling or working near chemical substances. Proper ventilation should be ensured in enclosed spaces to minimize the risk of inhalation. Spreader equipment should be thoroughly cleaned and decontaminated after use to prevent residual chemical exposure.
4. **Equipment Overturns**: Spreader equipment, especially when towed behind a vehicle, can be prone to overturns, particularly on uneven terrain or steep slopes.
   - Mitigation: Operators should adhere to recommended speed

limits and avoid operating spreaders on excessively steep or uneven terrain. Spreader loads should be distributed evenly to maintain stability, and operators should be trained in safe towing practices, including proper hitching techniques and load management.

5. **Entanglement Hazards**: Spreader equipment with moving parts such as conveyor belts or chains can pose entanglement hazards if loose clothing or body parts come into contact with them.
    - Mitigation: Operators should avoid wearing loose-fitting clothing or jewellery that could become caught in moving parts. Spreader equipment should be turned off and secured before performing any maintenance or adjustment tasks to prevent accidental entanglement.
6. **Electrical Hazards**: Electrically powered spreaders pose a risk of electrical shocks or fires if not properly maintained or operated.
    - Mitigation: Spreader equipment should be inspected regularly for damaged wiring or components and repaired or replaced as needed. Electrical connections should be properly insulated and protected from moisture to prevent short circuits. Operators should receive training on safe electrical practices and be aware of emergency shutdown procedures in case of electrical malfunctions.

By identifying these hazards and implementing appropriate mitigation strategies, operators can minimize the risks associated with spreader operations and ensure a safe working environment for themselves and others involved. Regular maintenance, proper training, and adherence to safety protocols are key to preventing accidents and injuries during spreader operations.

Planning and preparing for spreader operations, whether for fertilizers, manure, or other materials, is essential to ensure safe and efficient work practices. Here's a guide on how to plan and prepare for spreader operations, including pre-start checks:

1. **Assess the Site and Conditions**:
   - Before starting any spreader operations, assess the site and terrain conditions where the spreading will take place. Consider factors such as slope, soil type, obstacles, and weather conditions.
   - Identify any hazards or potential risks that could affect the safety and effectiveness of the operation, such as overhead power lines, uneven terrain, or nearby water bodies.
   - Plan the spreading routes and areas to be covered, taking into account the desired spread pattern and application rates.
2. **Select the Appropriate Spreader and Materials**:
   - Choose the right type of spreader for the specific task and materials to be spread. Consider factors such as the type of material (e.g., granular, liquid, or solid), application rates, and spreading width.
   - Ensure that the selected spreader is in good working condition and properly calibrated for the intended application. Refer to manufacturer guidelines and specifications for proper setup and operation.
3. **Prepare the Equipment**:
   - Conduct pre-start checks and inspections of the spreader

equipment to ensure it is safe and ready for use. This includes checking for any signs of damage, wear, or leaks.

- Inspect all moving parts, hydraulic systems, electrical connections, and safety features to ensure they are functioning correctly.
- Check fluid levels, such as fuel, hydraulic fluid, and lubricants, and top up as needed.
- Make sure all safety guards and shields are in place and properly secured to prevent accidents.

4. **Prepare the Materials**:
    - If using granular fertilizers or other dry materials, ensure that they are properly stored and handled according to manufacturer recommendations. Avoid moisture contamination or clumping, which can affect spreading accuracy.
    - For liquid or slurry applications, ensure that the materials are properly mixed and stored in appropriate containers or tanks. Check for any sediment or impurities that could clog the spreader equipment.
5. **Review Safety Procedures and Protocols**:
    - Review and communicate safety procedures and protocols with all personnel involved in the spreading operation. Emphasize the importance of following safe work practices and using appropriate personal protective equipment (PPE).
    - Establish emergency response protocols and ensure that all personnel are familiar with them. Identify emergency shutdown procedures, evacuation routes, and emergency contact

information.

6. **Perform Pre-Start Checks**:

    - Before starting the spreader operation, perform pre-start checks according to manufacturer guidelines and operator manuals.
    - Check all controls, gauges, and indicators to ensure they are functioning properly.
    - Test the spreading mechanism and adjust settings as needed to achieve the desired spread pattern and application rates.
    - Verify that all safety features, including emergency stop controls and warning lights, are operational.

By carefully planning and preparing for spreader operations, including thorough pre-start checks, operators can minimize risks, ensure equipment reliability, and achieve optimal results in their spreading tasks. Regular maintenance, adherence to safety protocols, and proper training are essential for safe and effective spreader operations.

Operating a Spreader

While the chemical composition of fertilizers is regulated, the physical characteristics such as particle size, shape, strength, and density are not standardized. These physical properties significantly affect the spreading process and the optimal adjustment of the spreader (Teagasc, 2023).

An ideal fertilizer typically comprises particles with a relatively large size, with about 80% falling within the 2mm to 4mm range. A simple sieve box can be used to check this. Most fertilizers have a density of around 1kg/litre, but urea, both standard and protected, has a lower density of about 0.8kg/litre, making it more challenging to spread. This necessitates different spreader settings and often limits the bout widths.

Larger particles facilitate spreading, especially in the case of low-density urea. However, the marketing of blends containing both low-density urea and high-density phosphorous (P) and potassium (K) elements poses a challenge as these constituents may spread to varying widths.

Particle strength is another crucial factor as weak particles can break upon hitting the spreader disc vane. Always inquire about the physical quality of the fertilizer from your supplier.

Different makes and models of fertilizer spreaders have varying capabilities concerning the bout width at which they can evenly spread different products. In most cases, urea may not be spread as widely or evenly.

Figure 94: Spreading liquid manure on a harvested corn field. Chesapeake Bay Program, CC BY 2.0, via Wikimedia Commons.

The most critical aspect of any spreader is its ability to evenly distribute fertilizer, typically determined through a tray test. This test produces a coefficient of variation (CV) value, with values below 10% indicating even spreading. The design of the disc, vane, and spreader outlet influences the spreading uniformity, which may not necessarily be improved by electronic control or GPS systems. It is insufficient to rely solely on the absence of striping in the target crop as evidence of even spreading—always seek evenness test data when purchasing a spreader.

Modern fertilizer spreaders usually feature adjustable spread patterns that can be modified to suit the bout width for the fertilizer product being used, considering its size, density, and strength. Changes may involve altering the disc type, vane type, vane position, spreader tilt, and fertilizer drop point, among others. Manufacturers often test a wide range of products to enable users to match their fertilizer with entries in the fertilizer manufacturer's database. Increasingly, this process is facilitated through smartphone apps or online platforms. Typically, a simple fertilizer sieve box, density check, and strength test are utilized to characterize the fertilizer and determine the appropriate settings. This should be done for every batch of fertilizer.

Setting the desired application rate can be aided by using resources provided by the spreader manufacturer, but static calibration using the actual fertilizer being applied is generally recommended. Some manufacturers offer calibration aids such as fertilizer flow rate calculators to simplify the task. Spreaders equipped with weigh cells or electronic flow measurement devices allow for self-calibration. Given that fertilizer varies from batch to batch, calibration checks are essential. Relying solely on estimates and making adjustments after completing each paddock or field can lead to incorrect application rates over a significant area.

Marking bout widths using tramlines in cereals, temporary markers, or GPS guidance is crucial for spreaders operating with bouts of 12m and more. While simple light bar or screen guidance GPS systems depend on operator skill for accuracy, higher-resolution systems and auto-steer can enhance accuracy at a considerable cost. GPS systems that automatically turn fertiliser flow on and off at the headlands can also improve accuracy. Modern spreaders often have turn-on points located 15m to 20m or more from the end headland, which can be challenging to judge by eye, potentially resulting in significant overspreading when leaving the headland (Teagasc, 2023). If GPS switching at the headlands is not feasible, the correct turn-on points should be marked in the field to allow operators to calibrate their eye for this distance.

Broadcast spreaders require a different spreading arrangement for use on field boundaries to achieve the correct rate up to the boundary and avoid spreading past it. Modern spreaders are designed with a large overlap of fertilizer from adjacent bouts to ensure even spreading. However, on headlands, one side of the pattern must deliver the same quantity of fertilizer without overlap to the field edge. Achieving this is challenging, and manufacturers employ various methods such as deflectors, different discs, vanes, and drop points, though none are perfect and often require meticulous setting using trays. Inaccurate on/off switching and improperly adjusted headland mechanisms can result in significant errors at field headlands. Correct selection, setting, and use of fertilizer spreaders are imperative to ensure even spreading, demanding accurate information about fertilizer quality. Getting the fundamentals right is essential.

The correct adjustment of a spreading machine depends on the physical characteristics of the fertilizer. Factors such as particle size distribution, bulk density, and flow rate significantly impact the settings.

While spreader manufacturer tables provide guidance, they are based on tests conducted under ideal conditions in test stations. They offer information on how to set the spreader for optimal, even spreading at the desired application rate (kg/ha) and specific working width. However, it's important to note that these tables serve as a guide only.

To achieve ideal spreading conditions, it's recommended to conduct a tray test using your specific combination of spreader and fertilizer. Separate tests should be performed for different types of fertilizers, as some products may vary depending on their origin. Always refer to the directions on the fertilizer bag for additional guidance.

Machine Height: Ensure that the disc or spout on a mounted spreader is positioned at the correct height above the crop, as outlined in the manufacturer's guide. Additionally, ensure that the machine is mounted correctly.

Fertilizer Application Recommendations and Guidance:

Forward Speed: The speed at which you travel plays a crucial role in maintaining a consistent application rate. Always use the recommended gear for your machine, especially with machines relying on PTO or ground drive to operate the disc. Variations in speed can result in uneven spreading and rate alterations.

Bout Width: It's essential to drive at the correct working width. Refer to the manufacturer's handbook for information regarding the type of fertilizer being used. For full-width distributors, each bout should align with the previous one.

Headlands: Leaving sufficient space for two bouts around the field is customary. For lateral spread machines, it's crucial to align the 'switch-on' and 'switch-off' of fertilizer to avoid double-dosing or missing areas. Utilize the headland disc or border device provided with the machine for this purpose. Consider application restrictions on field margins and adjust the headland device accordingly.

Completing Spreader Operations

To conclude spreader operations, it's crucial to undertake several essential steps to ensure that equipment is properly maintained, stored, and environmental concerns are addressed.

Cease Fertilizer Distribution: Once the intended area has been adequately covered, it's important to halt the spreader from distributing fertilizer. Follow the manufacturer's instructions carefully to disengage the spreader mechanism and prevent any unintended discharge of fertilizer.

Clean Equipment: After usage, it's essential to thoroughly clean the spreader to eliminate any remaining fertilizer residue. This helps prevent corrosion and buildup that could affect the equipment's performance and lifespan. Pay close attention to cleaning the hopper, distribution mechanism, and all moving parts.

Inspect for Damage: Conduct a visual inspection of the spreader to identify any signs of damage or wear. Check for cracks, loose bolts, or worn-out components that may require repair or replacement. Promptly address any issues to maintain the efficiency and safety of the equipment.

Store Properly: Proper storage is crucial to protect the spreader from the elements and minimize corrosion. Store the equipment in a dry, sheltered location and ensure it's securely stored to prevent accidental damage or theft. Follow any specific storage instructions provided by the manufacturer.

Dispose of Unused Fertilizer: Dispose of any unused fertilizer in accordance with local regulations. Avoid dumping excess fertilizer on the ground or in waterways, as this can contribute to pollution and environmental harm.

Record Keeping: Maintain accurate records of fertilizer usage, application rates, and any adjustments made to the spreader settings. These records are valuable for future planning, troubleshooting, and compliance with regulatory requirements.

Environmental Considerations: Take environmental factors into account, such as buffer zones, sensitive areas, or restrictions on fertilizer application near water bodies. Adhere to best management practices to minimize environmental impact and ensure compliance with regulations.

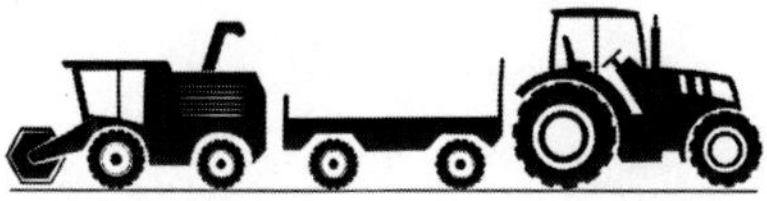

Chapter Thirteen

Quad Bikes

A quad bike, also known as an all-terrain vehicle (ATV), is a motorized vehicle designed for off-road use. It typically has four wheels, low-pressure tyres, and a straddle seating position for the operator. Quad bikes are commonly used for recreational purposes such as off-road trail riding, as well as for practical purposes such as agricultural work, forestry, and search and rescue operations. They come in various sizes and engine capacities, ranging from small models suitable for children to larger, more powerful machines for adults. Quad bikes are popular for their versatility and ability to traverse rough terrain. However, they can also be dangerous if not operated safely, so proper training and adherence to safety guidelines are essential for users.

Figure 95: Typical farm quad bike. Walter Baxter, CC BY-SA 2.0, via Wikimedia Commons.

Quad bikes are commonly used in agriculture for various tasks due to their versatility, agility, and ability to navigate rough terrain. Some of the ways quad bikes are used in agriculture include:

1. Farm Maintenance: Farmers use quad bikes to traverse their properties quickly and efficiently. They can inspect crops, check fences, and monitor livestock without the need for larger vehicles.

2. Crop Management: Quad bikes can be equipped with sprayers, spreaders, or other agricultural implements to apply fertilizers, pesticides, or herbicides to crops. Their manoeuvrability allows for precise application in areas where larger machinery may not reach.

3. Livestock Management: Farmers use quad bikes to check on

and manage livestock, including feeding, herding, and checking for signs of illness or injury. Quad bikes enable farmers to move swiftly between different areas of their property.

4. Hauling: Quad bikes equipped with trailers or cargo racks can transport small loads such as tools, feed, or harvested crops around the farm. They provide a convenient means of moving materials without the need for larger, more cumbersome vehicles.

5. Terrain Accessibility: Agricultural land can often have rough or uneven terrain that is challenging for larger vehicles to navigate. Quad bikes can access areas where traditional tractors or trucks cannot, making them invaluable for tasks in remote or hard-to-reach locations.

6. Emergency Response: In emergency situations such as fires or medical emergencies on the farm, quad bikes can provide rapid transportation for responders or supplies to reach the affected areas quickly.

Quad bikes play a vital role in modern agriculture by improving efficiency, productivity, and accessibility on farms of all sizes.

Figure 96: Farmer and dog on a quad bike. Karl and Ali, CC BY-SA 2.0, via Wikimedia Commons.

Many injuries or fatalities stem from rider inexperience, absence of helmets or other protective gear, and risky riding practices.

Despite being commonly referred to as all-terrain vehicles, quad bikes are not universally suitable for traversing all types of terrain. Despite their four-wheel configuration, quad bikes lack stability due to their high centre of gravity and narrow wheelbase. Most incidents resulting in injury or death involve the bike overturning onto the rider, often at low speeds.

Various factors contribute to injuries or fatalities involving quad bikes, such as failure to adhere to manufacturer instructions, entrapment of limbs by the tyres, overturning while navigating steep slopes, collisions with obstacles leading to rollovers, impacts from low-hanging obstacles like branches, overloading or improper distribution of weight, lack of familiarity with the vehicle's controls and handling, rider inexperience, reckless riding behaviours like excessive speed or attempting stunts, and inadequate maintenance leading to mechanical failures, particularly of crucial safety components like brakes.

To mitigate the risks associated with quad bike operation, it's crucial to familiarize oneself with the capabilities and limitations of the vehicle. This includes reading the manual thoroughly, prioritizing the use of approved safety gear like helmets meeting relevant standards, heeding manufacturers' safety warnings, ensuring all operators watch and comprehend safety instructional videos provided with the quad bike, seeking recommendations for training courses from relevant suppliers or institutions and agricultural colleges, and diligently practicing riding until confidence is attained before engaging in intended activities with the vehicle.

Planning and Preparing for Quad Bike Operations

Planning and preparing for quad bike riding on a farm is crucial for ensuring safety and efficiency. This includes:

Assess Riding Area: Before riding, survey the farm area where you intend to ride the quad bike. Identify potential hazards such as uneven terrain, obstacles, livestock, and other farm machinery. Determine the best routes to take based on the task at hand and the condition of the terrain.

Check Weather Conditions: Monitor weather forecasts to ensure safe riding conditions. Avoid riding in adverse weather conditions such as heavy rain, strong winds, or poor visibility, as these can increase the risk of accidents.

Inspect Quad Bike: Conduct a thorough inspection of the quad bike before each ride. Check for any signs of damage, loose bolts, leaks, or other mechanical issues. Ensure that all controls, brakes, lights, and tyres are in proper working condition.

Wear Proper Safety Gear: Before riding, ensure you and anyone else riding the quad bike wear appropriate safety gear. This includes a helmet, goggles or a face shield, gloves, long pants, and sturdy footwear. Ensure that all safety gear meets relevant safety standards.

Secure Load and Attachments: If you'll be carrying any loads or using attachments on the quad bike, ensure they are securely strapped down and properly balanced. Follow manufacturer guidelines for attaching and securing loads to prevent accidents or loss of control.

Inform Others: Let someone else know your riding plans, including where you'll be riding and when you expect to return. This is especially important if riding alone, as it ensures that someone is aware of your whereabouts in case of an emergency.

Carry Necessary Tools and Supplies: Carry a basic toolkit and any necessary supplies with you, such as a first aid kit, water, and a mobile phone for communication in case of emergencies.

Consider Training: If you're new to quad bike riding or unfamiliar with the specific quad bike you'll be using, consider undergoing training from a certified instructor. Training can help improve your riding skills and safety awareness.

Plan Routes and Tasks: Plan your routes and tasks carefully to minimize risks and maximize efficiency. Consider factors such as terrain, distance, time constraints, and the nature of the work to be done.

Follow Safety Guidelines: Always adhere to safety guidelines and regulations when riding a quad bike. Avoid risky behaviours such as excessive speed, carrying passengers, or performing stunts. Prioritize safety at all times.

Keep in mind that working alone on a quad bike poses unique risks, as there may be no immediate assistance available in case of an emergency. To mitigate these risks, several strategies should be applied:

1. Risk Assessment: Conduct a thorough risk assessment before starting work. Identify potential hazards such as uneven terrain,

obstacles, adverse weather conditions, and the presence of other vehicles or livestock.

2. Communication Plan: Establish a communication plan to ensure that someone is aware of your whereabouts and activities. This may include regularly checking in with a colleague, supervisor, or family member, especially when entering remote areas with limited cellular coverage.

3. Emergency Response Plan: Develop an emergency response plan outlining the steps to take in case of an accident, injury, or breakdown. This plan should include procedures for summoning help, administering first aid, and securing the quad bike to prevent further accidents.

4. Personal Protective Equipment (PPE): Wear appropriate PPE, including a helmet, goggles or face shield, gloves, long pants, and sturdy footwear, to protect yourself from potential injuries in the event of a fall or collision.

5. Training and Competency: Ensure that you are adequately trained and competent to operate the quad bike safely. This includes understanding the vehicle's controls, practicing safe riding techniques, and undergoing training in emergency procedures.

6. Regular Maintenance: Maintain the quad bike in good working condition by conducting regular inspections and servicing as per the manufacturer's recommendations. This helps prevent mechanical failures and reduces the risk of breakdowns while working alone.

7. Carry Essential Supplies: Carry essential supplies such as a first aid kit, water, snacks, a mobile phone or two-way radio, a map

or GPS device, and basic tools for minor repairs. These supplies can help you manage minor emergencies and ensure your well-being while working alone.

8. Limit Risky Behaviours: Avoid engaging in risky behaviours such as speeding, performing stunts, or riding in hazardous conditions. Stick to established safety protocols and operate the quad bike within your skill level and comfort zone.

9. Regular Check-ins: Schedule regular check-ins with a designated contact person to provide updates on your progress and well-being. This allows for timely intervention in case of any unforeseen incidents.

10. Weather Monitoring: Keep a close eye on weather forecasts and avoid riding in adverse weather conditions such as heavy rain, strong winds, or extreme temperatures, which can increase the risk of accidents or exposure-related illnesses.

By implementing these risk mitigation strategies, individuals working alone on a quad bike can minimize the likelihood of accidents and injuries, ensuring a safer and more productive working environment.

Operating a Quad Bike

The controls on a quad bike typically consist of various components that enable the rider to operate the vehicle safely and efficiently. Here's an explanation of the common controls found on quad bikes:

1. Throttle: The throttle control, usually located on the right-hand grip, regulates the engine speed. By twisting the throttle grip towards the rider, the engine accelerates, while releasing it slows down the engine.

2. Brakes: Quad bikes typically have two independent brake controls: the hand brake and the foot brake. The hand brake lever, usually located on the right-hand grip, operates the front brakes. The foot brake pedal, located on the right footrest, controls the rear brakes. Applying both brakes simultaneously ensures effective braking.

3. Gear Shifter: Most quad bikes have a foot-operated gear shifter located on the left footrest. It allows the rider to shift between gears (forward, neutral, and reverse) to control the direction of travel.

4. Ignition Switch: The ignition switch, usually located near the handlebars or on the instrument panel, activates the quad bike's electrical system and starts the engine.

5. Lights and Indicators: Controls for headlights, taillights, turn signals, and other lighting features are typically located on the handlebars or dashboard. They enable the rider to illuminate the path and signal intentions to other road users.

6. Kill Switch: The kill switch, often located on the handlebars, is an emergency shut-off mechanism that immediately stops the engine when activated. It's used in situations where rapid shutdown is necessary, such as in emergencies or if the rider loses control of the vehicle.

7. Ignition Key: Some quad bikes are equipped with an ignition key, which serves as an additional security measure to prevent unauthorized use of the vehicle.

8. Speed Limiter: Some quad bikes may feature a speed limiter control, allowing the rider to set a maximum speed limit for safety, particularly useful for beginners or when riding in restricted

areas.

These controls may vary slightly depending on the make and model of the quad bike, but they generally serve the same basic functions. It's important for riders to familiarize themselves with the location and operation of these controls before operating a quad bike to ensure safe and effective use of the vehicle.

Riding a quad bike safely and effectively involves several key steps and considerations. First and foremost, before setting out, it's crucial to ensure you're wearing appropriate safety gear. This includes a helmet, goggles or a face shield, gloves, long pants, and sturdy footwear. Additionally, take the time to familiarize yourself with the quad bike's controls, operation manual, and any safety features it may have.

When mounting the quad bike, approach it from the left side for stability. Grasp the handlebars firmly, swing your leg over the seat, and settle into the seating position. Ensure both feet are securely positioned on the footrests.

Starting the engine is the next step. Insert the ignition key and turn it to the "on" position. Follow the specific starting procedure outlined in the owner's manual, which may involve pulling in the clutch (if equipped), pressing the ignition switch, and starting the engine.

Once the engine is running, familiarize yourself with the throttle control located on the right-hand grip. To accelerate, twist the throttle towards you; to decelerate, release it. Use the gear shifter, typically located on the left footrest, to select forward, neutral, or reverse gears as needed.

Understanding and becoming acquainted with both the hand brake and foot brake controls is essential for effective braking. The hand brake, located on the right-hand grip, operates the front brakes, while the foot brake, located on the right footrest, controls the rear brakes. Using both brakes simultaneously ensures efficient braking.

When it comes to riding techniques, start in an open, flat area to practice. Keep your body centred and balanced on the quad bike, with your feet on the footrests and hands firmly gripping the handlebars. Anticipate obstacles or changes in terrain and adjust your riding accordingly.

Turning the quad bike involves leaning your body slightly in the direction of the turn while maintaining a firm grip on the handlebars. Avoid making sudden or sharp turns, especially at high speeds.

Maintain awareness of the terrain you're riding on and adjust your speed and riding technique accordingly. Be cautious when riding on uneven, rocky, or slippery surfaces, and avoid steep inclines or obstacles whenever possible.

Always prioritize safety by riding at a safe and controlled speed, obeying all traffic laws and regulations. Avoid reckless manoeuvres, such as wheelies or excessive speeding, which can increase the risk of accidents.

When it's time to dismount, ensure the quad bike is in neutral and apply the brakes. Swing your leg back over the seat and dismount from the left side of the quad bike.

Remember, practice and familiarity with your quad bike are essential for safe and enjoyable riding experiences. If you're unsure about any aspect of riding, seek guidance from experienced riders or professional instructors.

Farmers are advised to install an operator protective device (OPD) specifically designed and tested for quad bike rollover risks. Additional safety guidelines for quad bikes include:

- Adhere strictly to the manufacturer's instructions for operating the quad bike.
- Maintain all safety guards in their designated positions.
- When equipping the quad bike with accessories, ensure they are

either manufacturer-approved or recommended.

- Properly install accessories without modifying their fit to avoid compromising the quad bike's stability.
- Adhere strictly to load ratings.
- Keep the quad bike well-maintained mechanically.
- Conduct a safety inspection before each ride.
- Avoid traversing steep inclines.
- Stay within terrain that matches your riding proficiency level.

Figure 97: New Holland tractor and quad bike with OPD. Evelyn Simak, CC BY-SA 2.0, via Wikimedia Commons.

Suggestions for riders encompass:

- Treat the quad bike as industrial machinery, not a recreational vehicle, avoiding irresponsible riding or attempting stunts like wheelies.
- Restrict quad bike operation to trained individuals only.
- Educate children about potential hazards and prohibit their use of the quad bike until they receive proper training and supervision, ensuring they remain at a safe distance when others are riding.

- Refrain from carrying passengers on the quad bike, as it limits the rider's ability to adjust weight distribution appropriately.

- Always wear suitable protective gear, including an standards-compliant helmet (goggles if the helmet lacks a visor), boots, gloves, durable trousers, and a jacket.

- Maintain an appropriate speed at all times and reduce speed before turning or braking.

Terrain-related safety tips include:

- Opt for familiar tracks whenever possible, considering factors like obstacles, weather conditions, surface characteristics, and required speed.

- Exercise caution when riding on bitumen roads due to potentially compromised control on smooth surfaces.

- Be aware of the shifting centre of gravity caused by liquids in spray tanks, ensuring additional weight does not exceed the carrying capacity to prevent accidents.

- Evaluate terrain carefully, avoiding steep slopes, especially when loose or wet dirt increases rollover risks.

- Scan the ground ahead for potential hazards like rocks or pipes to prevent accidents.

- If uncertain about navigating specific terrain, opt for an alternative route or turn back to mitigate risks.

To start the ATV, insert the key into the ignition and turn it to the start position. Then, press the start button, typically located on the right side of the handlebars. Once the engine starts, allow it to idle for approximately a minute to warm up. In colder weather conditions, it's advisable to let the engine warm up for about 5 minutes before riding.

To engage the engine into neutral, pull the clutch handle, located on the left handlebar. This action disengages the engine from the gears, allowing you to shift gears smoothly as you accelerate. To begin moving, release the clutch gradually with your left hand to engage the engine into gear. While in neutral, the ATV can roll forward, but no speed can be added. Transitioning into first gear is necessary to initiate movement.

Use your left foot to raise the gear shift lever, found on the left footrest, to shift into higher gears. With the clutch engaged, lift the lever to shift gears and then release the clutch to engage the engine. As you gain speed, shift into higher gears gradually. Practice riding at various speeds and shifting into higher gears to become accustomed to the process.

For ATVs equipped with automatic transmission, shifting gears is unnecessary. Simply focus on gradually increasing speed to become more comfortable with riding.

When slowing down, downshift into lower gears to match the reduced speed. Hold the clutch with your left hand and press down on the gear shift lever with your left foot, then release the clutch. Each downshift should be done individually to allow the engine to adjust to lower speeds and gears.

To apply brakes, start by using your right hand to engage the rear brakes, followed by gradually adding pressure with your left hand to engage the front brakes. Applying both brakes simultaneously may cause you to tumble forward over the handlebars, while solely using the front brake may result in the ATV flipping over.

Maintain stability during turns by leaning into them. Shift your weight towards the direction of the turn to distribute weight evenly and prevent tipping. Practice leaning into turns, and if necessary, stand up from the seat to lean further, particularly during sharper turns.

For challenging terrain such as rocky, steep, uneven, sandy, or muddy areas, it's important to be vigilant. Look out for potential hazards such as rocks, holes, fallen branches, and uneven ground. If visibility of ground conditions is poor, consider dismounting and walking the area before driving over it. Alternatively, utilizing a drone may provide assistance in assessing terrain conditions.

Carrying passengers on quad bikes designed for single riders can significantly increase instability by raising the centre of gravity and restricting the rider's ability to use active riding techniques. Therefore, it's advised to never carry passengers on a quad bike designed for solo use.

When towing with a quad bike, it's crucial to adhere to load limits stated in the owner's manual. Using a trailer that is too heavy, large, or has an incompatible centre of gravity can lead to jack-knifing, loss of traction, or rollovers.

When carrying loads, particularly heavy or unstable ones like chemicals for spraying, maintaining balance and keeping the load low is essential. Ensure that the load is securely strapped down and consider using internal baffles in tanks for liquids to prevent alterations in the vehicle's centre of gravity. It's also imperative to follow load limits outlined in the owner's manual.

Maintaining proper tyre pressure is key to ensuring traction and stability. Check tyre pressure before start-up and during use, following the instructions provided in the owner's manual.

Finally, when considering attachments for your quad bike, ensure they are suitable for use and follow attachment instructions provided in the owner's manual. Using attachments that are not compatible can reduce stability, operator control, and overall performance of the vehicle.

Completing Quad Bike Operations

Completion and Shutdown is a critical phase in quad bike operations, ensuring that the vehicle is properly cared for and safely stored after completing tasks. This involves:

1. Once the tasks are completed, return the quad bike to a safe location: After finishing the required tasks, it's essential to return the quad bike to a designated safe area. This area should be free from potential hazards such as moving machinery, uneven terrain, or obstructions. By returning the quad bike to a safe location, the risk of accidents or damage to the vehicle is minimized.

2. Turn off the engine by switching the ignition key to the "off" position: To prevent unnecessary fuel consumption and reduce the risk of accidents, it's crucial to turn off the engine of the quad bike. This is done by switching the ignition key to the "off" position. By cutting off the engine's power supply, the quad bike is effectively shut down, ensuring safety and conserving fuel for future use.

3. Conduct a post-ride inspection to check for any damage or issues with the quad bike: After shutting down the quad bike, it's important to conduct a thorough post-ride inspection to identify any damage or issues that may have occurred during the operation. This inspection should include checking for signs of wear and tear, loose bolts or fittings, fluid leaks, tyre pressure, and any other potential safety concerns. Identifying and addressing issues promptly can prevent accidents and prolong the lifespan of the quad bike.

4. Properly store the quad bike in a designated area, ensuring it is secure and out of the way of other operations: Once the

post-ride inspection is complete and any necessary repairs or maintenance tasks are addressed, the quad bike should be stored in a designated area. This area should be secure, protecting the quad bike from theft, vandalism, or environmental damage. Additionally, the quad bike should be stored in a location that does not obstruct other operations on the farm or pose a safety risk to workers or equipment.

The Completion and Shutdown phase of quad bike operations involves returning the vehicle to a safe location, turning off the engine, conducting a post-ride inspection, and properly storing the quad bike to ensure its safety and longevity. By following these steps, quad bike operators can maintain the vehicle's reliability and minimize the risk of accidents or damage.

CHAPTER FOURTEEN

Biosecurity in Agricultural Settings

Biosecurity encompasses preventative measures aimed at minimizing the transmission of infectious diseases, invasive pests, or weeds. Effective biosecurity practices not only prevent the spread of infectious diseases and invasive species between farms but also safeguard the locality from diseases and weeds originating overseas. These measures also include protocols for containing disease outbreaks when they occur.

Farm-specific equipment and machinery are preferable to limit the entry of items shared with other sites. Any shared items should undergo thorough cleaning and disinfection before crossing the line of separation (LOS). Heavy machinery like tractors or skid loaders used in animal areas tend to accumulate dirt, potentially spreading germs within the farm and to other farms. It's essential to clean and disinfect them before moving them to other animal areas.

Planting materials and other farm inputs can introduce pests onto the farm. While certified planting material helps reduce risks, all farm inputs should undergo close inspection to prevent pest introduction. Poor-quality or infected planting material poses a threat to the business. Any suspicion of pests or unhealthy planting material entering or leaving the farm should be promptly reported to the farm manager or supervisor.

Transporting farm produce within and between farms carries the risk of spreading pests. Before transportation, it's important to inspect produce for pests or unusual symptoms. Additionally, ensure compliance with people and vehicle movement recommendations to minimize the risk of spreading pests during transportation.

Any pests or unusual symptoms observed in harvested produce, or any transport activities compromising farm biosecurity, should be reported immediately to the farm manager or supervisor. Prompt reporting is crucial for implementing appropriate measures to mitigate the risk of pest spread and maintain biosecurity standards on the farm.

Biosecurity in relation to agricultural equipment refers to the measures and practices implemented to prevent the introduction and spread of pests, diseases, and pathogens through equipment used in agricultural operations. This is crucial for protecting crops, livestock, and the environment from potential threats that can have significant economic and ecological consequences. Here's an explanation of biosecurity principles as they relate to agricultural equipment:

1. **Equipment Cleaning and Disinfection**:

    - Regular cleaning and disinfection of agricultural equipment, such as tractors, harvesters, and tillage implements, are essential to remove any organic material, soil, or debris that may harbor pests, diseases, or pathogens. Disinfectants specifically formulated to target agricultural pathogens should be used to ensure thorough sanitization.

2. **Quarantine Procedures**:

- New or imported agricultural equipment should undergo quarantine procedures before being introduced to farm operations. This may involve isolating the equipment in designated areas and conducting thorough inspections to detect and address any potential biosecurity risks before use.

3. **Equipment Inspection and Maintenance**:
   - Routine inspection and maintenance of agricultural equipment are necessary to identify and address any issues that may compromise biosecurity. This includes checking for leaks, damaged components, or areas where pests or pathogens could accumulate and spread.
4. **Restricted Access**:
   - Restricting access to agricultural equipment to authorized personnel only helps prevent unauthorized individuals from inadvertently introducing pests, diseases, or pathogens to farm premises. Secure storage areas for equipment when not in use can further minimize the risk of contamination.
5. **Training and Awareness**:
   - Educating farm workers and equipment operators about the importance of biosecurity and providing training on proper equipment handling, cleaning, and disinfection procedures is vital. Increased awareness among personnel helps ensure compliance with biosecurity protocols and reduces the risk of unintentional spread of pests or diseases.
6. **Record Keeping**:
   - Maintaining detailed records of equipment usage, cleaning, and disinfection activities can aid in traceability and ac-

countability. This information facilitates the identification of potential biosecurity breaches and allows for swift corrective action to prevent further spread of pests or diseases.

7. **Biosecurity Zones**:

    - Implementing biosecurity zones within farm premises, especially in areas where high-value crops or livestock are present, helps control the movement of equipment and personnel to minimize the risk of cross-contamination. These zones may include designated entry and exit points, handwashing stations, and equipment decontamination areas.

By integrating these biosecurity measures into agricultural practices, farmers can mitigate the risk of introducing and spreading pests, diseases, and pathogens through equipment, safeguarding the health and productivity of crops, livestock, and the agricultural environment.

Biosecurity protocols concerning people, vehicles, and equipment on your farm are crucial for preventing the spread of diseases, pests, and weeds. The primary defence against associated risks involves restricting access and closely monitoring movements. This can be achieved by delineating restricted areas and controlling entry points.

Establish a clear workflow for regular service providers and ensure they access only the necessary areas. Accompany visitors at all times and designate an area for cleaning boots and equipment upon arrival and departure. Minimize equipment lending between farms and thoroughly clean them before use.

Visitors, including suppliers, veterinarians, and consultants, pose a risk of unintentionally carrying pests and diseases onto the property. Limiting entry points, directing visitors to designated parking areas, and requiring them to sign a register can help monitor movements. It's essential to ensure cleanliness of vehicles, equipment, boots, and clothing to prevent the introduction of pests and diseases.

Vehicles can serve as carriers for diseases, pests, and weeds, either directly or through soil, plant material, or manure. Maintain vehicle hygiene by establishing designated parking areas away from livestock or crops. Implement a high-pressure wash-down facility and regularly inspect the area for new pests or weeds. Keep records of equipment and vehicle cleaning and disinfect all borrowed or second-hand machinery before use.

Limit the entry of vehicles used for waste removal and ensure they are cleaned before entering the property. Store waste materials away from animal areas and provide separate storage for items like manure spreaders. If sharing animal transport vehicles, clean and disinfect them between uses. Additionally, clean dirty vehicle tyres and wheel wells before crossing any points of entry.

Signage plays a crucial role in communicating your property's biosecurity status and expectations to visitors. It's essential to not assume that visitors are aware of the necessary biosecurity measures for your property. To ensure the effectiveness of signage:

Ensure that signs are clear, visible, and well-maintained to effectively convey the intended message. This includes regular inspection and replacement of signs as needed.

Signs should contain simple and concise messages, such as "Do not enter the farm without prior approval" or "Use wash-down facilities for cleaning vehicles and machinery." Clear and straightforward signage helps visitors understand what is expected of them in terms of biosecurity practices.

Support your signage with other biosecurity measures, such as restricted access points, to reinforce the importance of adhering to biosecurity protocols. By combining signage with physical barriers or access controls, you create a comprehensive approach to biosecurity management on your property.

Improper disposal of crop waste poses risks such as pest infestation and contamination of new crops. This waste encompasses various materials, including crop residues, prunings, and packing shed waste. To address this concern, farms implement specific procedures for the proper disposal of waste materials:

Crop residues and prunings are typically removed or ploughed into the soil immediately after harvest to facilitate decomposition and minimize pest breeding grounds.

Any waste generated from crops or packing sheds should be disposed of away from farm production areas and water sources. Disposal methods may include burying the waste, hot-composting, or transporting it to a designated waste management facility.

Crop waste intended for use as animal fodder must undergo treatment before transport to ensure the elimination of pests, thereby preventing the spread of infestations.

Loading of fruits and vegetables onto trucks should occur on a sealed pad located outside production areas to prevent soil and plant material from contaminating the produce.

Bins and storage areas must be free from soil, plant material, and pests to maintain cleanliness and prevent the spread of contamination. Regular cleaning and inspection are essential to uphold these standards and mitigate risks associated with crop waste management.

Cleaning and disinfecting equipment, tools, and machinery is essential for maintaining biosecurity within and between properties. This process involves removing all soil and plant material from the equipment and then disinfecting it with an appropriate solution. However, the approach to cleaning and disinfection may vary depending on the specific equipment, farm practices, and potential pest threats.

Before proceeding with cleaning and disinfection, it's advisable to consult with the farm manager or supervisor to determine if there are any specific procedures or recommended chemicals to use. In cases where an appropriate chemical disinfectant is not available, cleaning with hot soapy water can effectively eliminate many pests.

Alternatively, equipment can be cleaned by soaking it in a bleach solution containing 1% available chlorine for approximately 10 minutes. It's important to note that prolonged soaking in bleach may cause metal components to rust and damage clothing, so this method should be used with caution.

For tools such as pruners, dipping them in methylated spirits can serve as an effective disinfection method, killing many pests on contact. However, it's crucial to ensure proper ventilation when working with methylated spirits due to its flammable nature.

When dealing with machinery, it may be necessary to dismantle the equipment and clean each component according to farm-specific instructions. This ensures thorough cleaning and disinfection, minimizing the risk of pest transmission and maintaining biosecurity standards across the farm.

References

Bartok, J. (2024, 17/3/2024). Sprayers and Spray Application Techniques.

Boyce, B. (2021). Machinery guide: Types of tractors and categories. *AgDaily*, (5th May 2021).

Canadian Centre of Occupational Health and Safety. (2024). *Tractors*. Retrieved 15/3/2024 from

Carmody, P. (2009). *14. Windrowing and harvesting*. G. R. a. D. Corporation.

Deveau, J. (2017, 17/3/2024). How to Properly Set Up a Crop Sprayer.

Farmbrite. (2022, 15/3/2024). How to prepare a field for planting.

Hancock, D. W., Maddy, B., Gaskin, J., Baxter, L. L., & Harmon, D. D. (2022). *Preparing and Calibrating a No-Till or Conventional Drill for Establishing Forage or Cover Crops*.

Johnston, G. (2018). 10 Tips for better crop spraying. *Successful Farming*, (January 6, 2018).

Ozkan, H. E. (2016). *Best Management Practices for Boom Spraying*. O. S. U. Extension.

Rajput, P. (2023, 15/3/2024). Uses of Cultivator in Agriculture: Kheti-Gaadi.

Searle, S. (2018, 15/3/2024). What Cultivator Should You Use For The Job?

Smith, J. A., Schrock, M. D., Taylor, R. K., & Price, R. R. (2024). *Wheat Production and Pest Management for the Great Plains Region*. myFields.

Teagasc. (2023, 23/3/2024). Six steps you need to take for accurate fertiliser spreading.

Yazid. (2023, 15/3/2024). Top 5 Types of Cultivators To Increase Yields.

Index

Made in United States
Cleveland, OH
21 February 2025

14561245R00240